Scientific Research in World War II

This book seeks to explore how scientists across a number of countries managed to cope with the challenging circumstances created by World War II.

No scientist remained unaffected by the outbreak of World War II. As the book shows, there were basically two opposite ways in which the war encroached on the life of a scientific researcher. In some cases, the outbreak of the war led to engagement in research in support of a war-waging country; in the other extreme, it resulted in their marginalisation. The book, starting with the most marginalised scientist and ending with those fully engaged in the war effort, covers the whole spectrum of enormously varying scientific fates.

Distinctive features of the volume include:

- a focus on the experiences of 'ordinary' scientists, rather than on figureheads like Oppenheimer or Otto Hahn;
- contributions from a range of renowned academics including Mark Walker, an authority in the field of science in World War II;
- a detailed study of The Netherlands during the German Occupation.

This richly illustrated volume will be of major interest to researchers of the history of science, World War II and Modern History.

Ad Maas specialises in the history of physics in the nineteenth and twentieth centuries and in the Dutch and German scientific culture between 1750 and 1900. His published works have included a study on the 'Second Golden Age' of the Dutch sciences and the engineering activities of Albert Einstein.

Hans Hooijmaijers is Head of Collections at Museum Boerhaave, Leiden, The Netherlands. He has worked as a Researcher at the Max Planck Institut für Stromungsforschung, Göttingen, the University of Groningen, and Museum Boerhaave.

Routledge Studies in Modern History

1. Isolation
Places and practices of exclusion
Edited by Carolyn Strange and Alison Bashford

2. From Slave Trade to Empire
European colonisation of black Africa 1780s–1880s
Edited by Olivier Pétré Grenouilleau

3. Revolution, Counter-Revolution and Revisionism in Postcolonial Africa
The case of Mozambique, 1975–1994
Alice Dinerman

4. Charity and Mutual Aid in Europe and North America since 1800
Edited by Paul Bridgen and Bernard Harris

5. Churchill, Roosevelt and India
Propaganda during World War II
Auriol Weigold

6. Genocide and Fascism
The eliminationist drive in Fascist Europe
Aristotle Kallis

7. Scientific Research in World War II
What scientists did in the war
Edited by Ad Maas and Hans Hooijmaijers

Scientific Research in World War II

What scientists did in the war

Edited by
Ad Maas and Hans Hooijmaijers

Routledge
Taylor & Francis Group
LONDON AND NEW YORK

First published 2009
by Routledge
2 Park Square, Milton Park, Abingdon, Oxon OX14 4RN

Simultaneously published in the USA and Canada
by Routledge
270 Madison Avenue, New York, NY 10016

Routledge is an imprint of the Taylor & Francis Group, an informa business.

Typeset in Times New Roman by Saxon Graphics Ltd, Derby
Printed and bound in Great Britain by MPG Books Ltd, Bodmin

British Library Cataloguing in Publication Data
A catalogue record for this book is available from the British Library

Library of Congress Cataloging in Publication Data
Maas, Ad.
Scientific research in World War II : what scientists did in the war / Ad Maas and Hans Hooijmaijers.
p. cm. – (Routledge studies in modern history)
Includes bibliographical references and index.
ISBN 978-0-7103-1340-9 (hardback) – ISBN 978-0-203-88318-1 (ebook)
1. Scientists–History–20th century–Congresses. 2. World War, 1939–1945–Science–Congresses. 3. Research–History–Congresses. I. Hooijmaijers, Hans. II. Title.
Q141.H195 2009
509'.044–dc22
2008033118

ISBN 10: 0-7103-1340-3 (hbk)
ISBN 10: 0-203-88318-7 (ebk)

ISBN 13: 978-0-7103-1340-9 (hbk)
ISBN 13: 978-0-203-88318-1 (ebk)

Contents

Illustrations

Figures

Tables

Contributors

Chris C. Bissell is Professor of Telematics at the Open University, UK. He has contributed to distance teaching materials in electronics, control engineering, telecommunications and media studies, and authored two undergraduate textbooks. His major research interest is in the history of technology, particularly automation and control in Germany and the former Soviet Union, with over thirty refereed articles on the topic. Further information can be found at http://technology.open.ac.uk/tel/people/bissell/.

Open University Faculty of Technology, Walton Hall, Milton Keynes MK7 6AA, United Kingdom. c.c.bissel@open.ac.uk

Marlene Burns graduated as a mature student, with BA(Hons) History, through the Open University, UK, and followed this, in 2000, with an MA dissertation on the history of penicillin research in The Netherlands during the Second World War. She wrote a PhD on 'The Development of Penicillin in the Netherlands 1940–1950: The Pivotal Role of NV Nederlandsche Gist- en Spiritusfabriek, Delft' (Department of History, University of Sheffield, UK, September 2005). Currently, she is carrying out research at the Delft University of Technology, The Netherlands, on the dissemination of information about penicillin during the war years. She also hopes to publish her PhD thesis in book form soon.

The Kluyver Archive, The Delft School Archives, Department of Biotechnology, Delft University of Technology, Julianalaan 67, 2628BC, Delft, The Netherlands. M.Burns-Fitzpatrick@tudelft.nl

Dirk van Delft studied physics at Leiden University. His book *Freezing Physics: Heike Kamerlingh Onnes and the Quest for Cold* was published in 2007 (Edita). He is Director of Museum Boerhaave, Leiden, The Netherlands, the Dutch national museum of the history of science and medicine, and Extraordinary Professor in the History of Science at Leiden University, The Netherlands.

Museum Boerhaave, Postbus 11280, 2301 EG, Leiden, The Netherlands. dirkvandelft@museumboerhaave.nl

J. V. Field is a Visiting Research Fellow at Birkbeck, University of London, and works on the history of the mathematical sciences in the period from about 1400 to 1650, but was formerly a computer programmer and in that capacity helped to design the optical system of the Anglo-Australian telescope (Coonabarabran) using Cambridge University's Edsac II and Titan computers. Publications include *Kepler's Geometrical Cosmology* (Chicago: University of Chicago Press, 1988), *The Invention of Infinity: Mathematics and Art in the Renaissance* (New York: Oxford University Press, 1997) and *Piero della Francesca: A Mathematician's Art* (New Haven: Yale University Press, 2005).

School of History of Art, Film and Visual Media, Birkbeck, University of London, 43 Gordon Square, London WC1H 0PD, United Kingdom. jv.field@hart.bbk.ac.uk

Marian Fournier was, until her retirement early in 2007, Vice-Director of Museum Boerhaave, Leiden, The Netherlands. She was the museum's expert on microscopes and microscopy. Two of her major publications are *The Fabric of Life: Microscopy in the Seventeenth Century* (Baltimore: Johns Hopkins University Press, 1996) and *Early Microscopes: A Descriptive Catalogue* (Leiden: Museum Boerhaave, 2003). In the final years of her career she turned from the early microscopes to the twentieth-century electron microscope.

Museum Boerhaave, P.O. Box 11280, 2301 EG, Leiden, The Netherlands. research@museumboerhaave.nl

Leonardo Gariboldi is Research Fellow in History of Physics at the Istituto di Fisica Generale Applicata and at the Museo Astronomico-Orto Botanico di Brera of the Università degli Studi di Milano, Italy. His main fields of research are the history of cosmic ray physics (in particular, the biographies of Beppo Occhialini and Constance Dilworth), the history of physics in Milan and the history of scientific instruments. Among his publications are some of the historical chapters in *The Scientific Legacy of Beppo Occhialini*, edited by P. Redondi, G. Sironi, P. Tucci and G. Vegni (Berlin, Heidelberg: Springer Verlag, 2007).

Università degli Studi di Milano, Istituto di Fisica Generale Applicata, Sezione di Storia della Fisica, Via Brera, 28, 20121, Milano (MI), Italy. leonardo.gariboldi@unimi.it

Hans Hooijmaijers is Head of Collections at Museum Boerhaave, Leiden, The Netherlands. He studied physics in Nijmegen and continued as a Researcher at the Max Planck Institut für Stromungsforschung, Göttingen, and the University of Groningen. He began as a Curator of Physics and Astronomy, focussing on the period between 1600 and 1900. He organised several exhibitions on science in sports, meteorology, light, food, Christiaan Huygens and clocks. He wrote a descriptive catalogue of all clocks in the collection of the museum. In 2005 he organised an international conference on the users of scientific instruments and edited its proceedings.

Museum Boerhaave, P.O. Box 11280, 2301 EG, Leiden, The Netherlands. hanshooijmaijers@museumboerhaave.nl

Alexander von Lünen did his PhD at the University of Technology, Darmstadt, Germany. He specialised in the history of diving and aerospace medicine and wrote several publications on the subject. His thesis is dealing with the history of diving and aviator equipment, with a focus on pressure suit development in Europe in the interwar years. His principal research interests (and publications) include history of physiology, history of diving, aviation and space travel (technology, medicine and culture) and history and computing. In respect to the latter, he joined the 'Great Britain Historical Geographical Information System' as Senior Research Associate in August 2007, at the Geography Department of the University of Portsmouth.

Department of Geography, University of Portsmouth, Buckingham Building, Lion Terrace, Portsmouth PO1 3HE, United Kingdom. avl1@gmx.de

Ad Maas is Curator History at Museum Boerhaave, Leiden, The Netherlands. He specialised in the history of physics in the nineteenth and twentieth centuries and in the Dutch and German scientific culture between 1750 and 1900. He wrote a PhD thesis on the history of physics at the University of Amsterdam between 1877 and 1940, and has published, among other things, about the rise of the so-called 'Second Golden Age' of the Dutch sciences (the decades around 1900) and the engineering activities of Albert Einstein. He organised exhibitions on acoustics, the sun and on Einstein and The Netherlands.

Museum Boerhaave, Postbus 11280, 2301 EG, Leiden, The Netherlands. admaas@museumboerhaave.nl

Falk Müller finished his PhD in 2003 with a thesis on the history of gas discharge physics in the nineteenth century. The general focus of his research is the history of the physical sciences in the nineteenth and twentieth centuries, particularly the history of experiment and of industrial physics. Currently, he is Wissenschaftlicher Assistant and teaches history of science at the Johann Wolfgang Goethe University, Frankfurt, Germany. He is preparing a book on the history of electron microscopy in Germany.

Johann Wolfgang Goethe Universität, History of Science, Grüneburgplatz 1, D-60629 Frankfurt am Main, Germany. falk.mueller@em.uni-frankfurt.de

Keiko Nagase-Reimer is Postdoctoral Research Fellow of the Section of Japanese History, Faculty of East Asian Studies, Ruhr-University Bochum, Germany. Her publications include *Forschungen zur Nutzung der Kernenergie in Japan, 1938–1945* (Marburg: Marburger-Japan-Reihe Band 30, 2002).

Japanese History Section, Faculty of East Asian Studies, Ruhr-University Bochum, Universitätsstr. 150, 44780 Bochum, Germany. Keiko.Nagase-Reimer@rub.de

Florian Schmaltz is Postdoctoral Research Fellow of the Historical Department of the Goethe University, Frankfurt, Germany. He wrote his PhD within the Research Programme of the Max Planck Society: 'History of the Kaiser Wilhelm Society in the National Socialist Era' (2000–2004). The title of his PhD was 'Kampfstoff-Forschung im Nationalsozialismus: Zur Kooperation von Kaiser-Wilhelm-Instituten, Militär und Industrie' (2005). His current DFG research project concerns 'Aerodynamic Research during the First and Second World War between Politics, Armament and Scientific Theories'.

Goethe-Universität Frankfurt, Historical Department – History of Science, Grüneburgpark 1, D-60323 Frankfurt am Main, Germany. schmaltz@em.uni-frankfurt.de

Stephen Snelders is Senior Research Fellow in the History of Medicine at the VU-University Medical Center, Department of Medical Humanities (Metamedica), Amsterdam, The Netherlands. Among his research interests are the history of genetics and theories of human heredity (in particular their uses in society and medicine), the history of drugs and the role of ship's surgeons in the travelling of medical knowledge in the seventeenth and eighteenth centuries. He has written various books and articles on these topics. One of his current book projects is a biography of Dutch SS doctors and eugenicists in the Second World War.

VU-University Medical Centre Metamedica, PO Box 7057, 1007 MB Amsterdam, The Netherlands. s.snelders@vumc.nl

Mark Walker is the John Bigelow Professor of History at Union College, where he teaches modern European history, including the history of science and technology and intellectual history. He has published on the history of science and technology under National Socialism, including comparisons with other regimes. His most recent books include *Physiker zwischen Autonomie und Anpassung – Die DPG im Dritte Reich*, edited with Dieter Hoffmann (Weinheim: VCH, 2007) and *Politics and Science in Wartime: Comparative International Perspectives on the Kaiser Wilhelm Institutes* [Volume 20 of *Osiris*], edited with Carola Sachse (Chicago: University of Chicago Press, 2005).

Union College, Department of History, Social Sciences Building, 807 Union Street, Schenectady, NY 12308-3107, USA. walkerm@union.edu

Masakatsu Yamazaki is Professor of History of Science of the Tokyo Institute of Technology, Japan. He edited a book about the history of atomic bombs: *Genbaku wa koushite kaihatu sareta* [*Atomic Bombs Were Thus Developed*], with Shizue Hinokawa (Tokyo: Aoki Publisher, 1990, 1997). His current research topic is the history of the nuclear energy research and development in wartime and postwar Japan. He has written several publications on this subject.

Graduate School of Decision Science and Technology, Tokyo Institute of Technology, 2-12-1 O-okayama, Meguro-ku, Tokyo, 152-8554, Japan. yamazaki@me.titech.ac.jp

Introduction
Ordinary scientists in extraordinary circumstances

Ad Maas

This book is the result of the *Scientific Research in World War II* conference, which was organised by Museum Boerhaave, the Dutch national museum for the history of science and medicine, on 25 and 26 January 2007. The aim was to explore the influence of the extraordinary circumstances created by the war on scientists and their research. The result is this compilation of thirteen articles.

Traditionally, the history of science in World War II has often been pursued in the service of questions of a 'larger' meaning, such as the influence of science on the outcome of the war: why did the Germans fail to make an atomic bomb while the Americans succeeded? Why and to what extent did German science decline altogether – and did it really? Next, the history of World War II science has almost inevitably focused on moral judgements of the behaviour of scientists, most usually concentrating on figureheads such as Werner Heisenberg, Otto Hahn, Robert Oppenheimer and Werner von Braun. Additionally, historians have questioned how the German and Japanese scientific environments could lead to immoral behaviour by scientists concerning human experimentation, for instance. Other recurrent themes are the forced migration (in particular, of the Jews) and the political attacks on scientific autonomy in the Third Reich. In summary, the history of science during World War II has traditionally been mainly concerned with the course and outcome of the war and with moral questions.[1]

Here, we put the emphasis on the consequences for the sciences and the scientists themselves: what did the extraordinary circumstances created by the war mean for the scientist and his research? This approach has yielded a number of largely factual, one could even say down-to-earth, case studies concentrating on 'ordinary' scientists coping with the extraordinary circumstances of war. We think that this perspective corroborates modern historiography in which it has gradually been realised that the eye-catching events and scientists do not necessarily have to be representative of science as a whole. The project on the Kaiser Wilhelm Society in National Socialism, Roy M. MacLeod's collection on science in the Pacific, David Edgerton's remarkable reinterpretation of twentieth-century Great Britain ('Warfare State'), Ute Deichmann's surprising conclusions concerning the continuity of the German sciences in Nazi Germany, Margit Szöllösi-Janze's focus on less obvious sciences in the Third Reich and the international comparisons in the recent Osiris volume *Politics and Science in Wartime* (2005) all concentrate on less visible structures and people, resulting in new insights into what scientific research in World War II concerned and into the role of the scientists.[2]

No mention is made in this book of Robert Oppenheimer, Otto Hahn or Lise Meitner, nor will Niels Bohr or Werner von Braun enter the stage. Apart from this introduction, no visits are paid to Los Alamos or Peenemünde. Our figureheads are the Delft engineer Jan Bart le Poole and the Italian physicist Guiseppe Occhialini, our only Nobel Prizewinner Ernst Ruska. Our favourite themes are the electron microscope, penicillin and the ill-fated Japanese large cyclotron, the most popular country The Netherlands. Yet, it is precisely these stories that can help to reveal how 'ordinary' scientific researchers behaved during World War II, how they maintained themselves and what kind of work they produced.

The 12 case studies in this volume are preceded by a general survey of science in World War II by Mark Walker. In this article he points to the imbalances in historiography, some of which I have just mentioned, as well as the latest trends. He describes the war scientist as someone who was profoundly affected by the circumstances of war, as someone whose freedom was limited but who could also, once the war broke out, surprisingly acquire some autonomy and influence. Above all, the scientist's fate could vary enormously ('either feast or famine'). Walker portrays him (or, occasionally, her) as someone who slipped easily into immoral behaviour, but also points to the difficulties in giving a clear-cut moral judgement of this behaviour.

Science under pressure

It was hardly possible for a scientist to duck out of the influence of World War II. 'In total war', as Mark Walker writes in this volume, 'when whole societies or populations are mobilised for the needs of a modern, industrial and technological conflict, almost all scientific activity becomes war work'. Basically, there were two opposite ways in which the war encroached on the life of a scientific researcher. The outbreak of the war could mean that he was engaged on research directly or indirectly related to the war effort of a warring country. In the other extreme case it could result in his marginalisation. Scientists who were introduced into the weapons and intelligence programmes of warring countries or who worked in strategic industries were often able to conduct research under improved or even splendid material conditions. On the other hand, they mostly had little time for contemplation, sometimes had hardly any freedom to determine their research topics at will and were confronted with tight deadlines.

To be sure, not every scientist was suited to meeting the requirements. At Los Alamos the retired Seth Neddermeyer, who liked working alone and in a methodical manner, acted too much 'as though it were just a normal research situation' in his experiments aimed at creating an implosion-design bomb. Consequently, he was sidelined by an impatient Oppenheimer.[3] Indeed, freedom and independence are perhaps considered the most precious assets of the modern researcher and one would expect that their loss must have been experienced as almost unbearable. In contrast, however, quite a few scientists experienced their engagement on the war effort as the most exciting time of their lives. The sense of urgency, of doing something that really mattered and the sense of belonging to a team of like-minded individuals all trying to achieve that one certain goal created a thrill that was never again matched during the rest of their careers. Moreover, the nature of the work was often found to be interesting and challenging. Scientists, deprived of their (academic) freedom and working under pressure, managed to achieve remarkable results.[4]

In her detailed account of the codebreaking activities at Bletchley Park, England, J. V. Field addresses just such a group of successful scientists both working under pressure ('we were hanging on by our fingernails') and being well supported. A group of linguists, classicists, chess players and mathematicians together put their talents to breaking the encrypted messages of the German Army. Illustrative of both the strenuous efforts of the scientists and the considerable material resources is the development over a short period of time of, in particular, the cutting-edge electronic counting machine 'Robinson' and its massive successor 'Colossus' (which is sometimes regarded as the first computer).

Work at Bletchley Park is sometimes said to have seriously shortened the duration of the war. A well-known achievement of this group was the deciphering of the Enigma code. Field elaborates, on the basis of material that was declassified only in 2000, on the breaking of messages encrypted by the so-called 'Tunny machine', which was used in the communications between the highest commanders, including Hitler himself, and which contained valuable strategic information for the Allies.

Also of influence on the course of the war were the developments in feedback control. In anti-aircraft weapons in particular it proved essential to have a stable feedback mechanism that was able to modify the position of the loop of the weapon rapidly and accurately in order to intercept the target. In his chapter, Chris Bissell sets out how feedback control research was organised in several warring countries: the UK, the US, Germany and the USSR. By highlighting this less visible but important example of war science and technology, Bissell provides a relevant contribution to this book, even though his international comparative approach is somewhat exceptional in this collection of 'down-to-earth' case studies.

As in the case of Bletchley Park, the discipline of feedback control saw a coming together of scientists from different disciplines who, under the pressure of the circumstances of war, raised the scientific and technological level of the field. Remarkably, while feedback control research in the different countries was organised in sometimes entirely different ways, they all independently developed a good deal of the theory of feedback control that has come to be known as 'classical control'.

In Japan it was only from the summer of 1942 that scientists were forced to contribute to the development of advanced weaponry – considerably later than in other war-waging countries.[5] The articles by Keiko Nagase-Reimer and Masakatsu Yamazaki clearly illustrate the rupture caused by this change of policy. The principal figure in both articles is the nuclear physicist Yoshio Nishina. While

Nagase-Reimer focuses on the vicissitudes of Nishina's large cyclotron, the pricey apparatus that was to secure him his place as one of the leaders in nuclear physics, but which tragically ended up in the Gulf of Tokyo, Yamazaki follows the experiences of Nishina's student, Masa Takeuchi. This latter approach gives an interesting insight into the fundamental research under Nishina's direction, which continued relatively unabated until, in December 1942, Nishina quite suddenly ordered Takeuchi to stop his experiments and start working on the nuclear weapons programme. He kept worrying about what motivated Nishina until the end of his days.

Why did Nishina make an offer to the Japanese Army to engage in research that would lead to a nuclear bomb? Both authors conclude that Nishina's main goal was to build his large cyclotron, which was only possible if he could embed it in a war-effort context, so that he secured sufficient means and manpower.

Scientific institutes in occupied territories in Europe also sometimes contributed to the war effort, though the Germans were rather reserved in introducing them directly into their programmes.[6] In his article, Florian Schmaltz elaborates on the peculiar case of the Dutch aerodynamic research institute (NLL), which conducted research projects for the Germans that were often, directly or indirectly, connected with their war effort. The Germans, who (perhaps rightly) did not trust the Dutch, avoided giving more information than was necessary. The Dutch researchers once made calculations on the Junker Ju 52 aircraft without knowing that their research concerned this well-known bomber and transport aircraft. It is clear that the motivation of the researchers at the NLL was different from that of their fellow scientists at Bletchley Park. Rather than working strenuously to help the country win the war, they worked for their own maintenance. Because the Germans limited their interference in the early years of the war largely to control through 'indirect rule', the workers were not fully engaged on working for the war effort. There were, in conclusion, gradations in the impact that working for the war effort had on those involved.

Science on the margins

The consequences of being marginalised, which was usually the case when the research had no relevance for the war effort or when the researcher was not allowed to enter war-related work, could include isolation from colleagues (abroad) and scarcity of equipment, energy and even paper. Scientists sometimes had to operate in societies that came to a standstill or they could be expelled from their workplace or even their country. However, in contrast to the scientist working for the war effort, they at least had the freedom to conduct research at will (if they had the opportunity to do so).

One such marginalised scientist was the Italian physicist Giuseppe Occhialini. As set out by Leonardo Gariboldi, this opponent of the Italian fascist regime was involved in raising a cosmic ray research group in São Paulo when Brazil joined the allied forces in 1942. As a citizen of an enemy country, he resigned from his post and spent his time as a mountain guide. Later, Occhialini was able to go to

England. Despite his determination to contribute to the war effort, he was not admitted to war-related activities by the British until the end of the war. Occhialini, in short, spent the war in powerlessness.

Another illustration of a scientist who was doomed to idleness is Jacob Jongbloed, presented by Alexander von Lünen. Jongbloed, who had done excellent work in the field of aviation medicine, spent the war giving courses and apparently without undertaking any research activities at all. No appeal was ever made to him on the part of German colleagues to collaborate on their research, while German aviation medicine was all but in a flourishing state. Among other things, it suffered from being cut off from English and American developments in particular. As von Lünen contends, the German aviation physicians shielded their comfortable and prestigious positions from foreigners – like Jongbloed – and were able to maintain an image of scientific superiority. As the title of the article ironically suggests, 'Splendid Isolation' is above all a story of the consequences of being isolated. The Germans, deprived of contacts with the Allied world, in their turn relegated Jongbloed to a kind of double isolation.

Being marginalised did not necessarily mean that scientists were unable to do any research. In an article on World War II science in the Soviet Union, Eduard Kolchinski has stated that 'even without freedom science can succeed brilliantly, while having absolute freedom with no government support it dies'.[7] Some articles in this volume show that (outside the Soviet Union) this does not always hold true. On the contrary, some remarkable achievements were made in research that was not related to the war effort and was not supported by the government. An exemplary case is the development of Dutch penicillin, Bacinol, as demonstrated by Marlene Burns in her article. Researchers at the yeast factory NG&SF (Netherlands Yeast and Spirits Factory) in the city of Delft managed, in secret and completely on their own, and without access to the groundbreaking results of Florey and Chain in Oxford that formed the basis for British and American penicillin production, to develop exactly the same substance. The Delft researchers produced the first penicillin substance in August 1944 and in November 1945, six months after the liberation of The Netherlands, it was successfully administered for the first time.

In her article, Marian Fournier elaborates on another remarkable project in Delft. A small group of researchers, supported by the local Technical University and some Dutch companies (among which again the NG&SF), managed to make major improvements to the electron microscope. The principal figure was the young engineer Jan Bart le Poole who, despite wartime shortages and other inconveniences, devised techniques that enabled the electron microscope to change magnification continuously and to zoom in on certain selected areas. Using the inventions of le Poole, the Philips Company managed to become market leader in electron microscopy soon after the war.

It proved essential for the Delft electron microscope inventors, deprived as they were of contacts with the Allied world, that they could keep abreast of German developments. The leading German firms published widely on their electron microscopes and sent their scientists to meetings in order to promote their products. The qualities of the unassumingly designed Delft microscope were not appreciated by the Germans, who consequently showed no interest in it.

It has been noted by historians that despite the dramatic impact of 'total war', people often tended just to continue their daily routine as much as possible. The articles by Burns and Fournier and several other papers in this volume illustrate this inclination. In Dirk van Delft's narrative about the Kamerlingh Onnes Laboratory, the renowned low-temperature apparatus of the University of Leiden, continuity is the leitmotiv. Director Wander Johannes de Haas managed to keep the institute running, even when the rest of Leiden University was closed down. Until 1944, research continued more or less without interruption. Dozens of articles left the laboratory on topics so far removed from war reality as the magnetic properties of superconductive metals. But apart from these somewhat unworldly research activities, there were other things going on as well. De Haas was seduced, by an interesting amount of money, to perform services for the German intelligence service Cellastic, which caused him trouble after the war. Yet, at the same time, he hid from the Germans a stock of ten tons of uranium that he had purchased a year before the outbreak of the war. With its labyrinth of rooms and passages de Haas' laboratory became a hiding place for refugees and the home of illegal activities. What were de Haas' real intentions?

It is interesting that the fame of the Kamerlingh Onnes Laboratory appeared to be both a threat and a guarantee of its survival. On the one hand, it seduced the Germans to claim it or (parts of) its apparatus to deploy it for their war-related research. On the other, it was considered a cultural asset 'throughout the whole German Reich', which should then be left intact. At the end of the war the Germans took a limited number of instruments.

Science as usual

In her book *Biologen unter Hitler* (1992), Ute Deichmann has identified a surprising degree of continuity in biological research in the Third Reich. Those who were not expelled from the German scientific community for political or ideological reasons could often carry on their research without much difficulty. With some adaptation to the circumstances and by showing a willingness to work for the Nazi cause, it was generally possible to retain endowment. Biologists often quite easily managed to argue the value of their research to the war effort (such as Willem Goetsch in Breslau for his study on termites and other subtropical 'vermins'). Deichmann's biologists belonged neither to the scientists who strenuously contributed directly to the war effort, nor to those who became marginalised.

This collection yields two other types of scientists belonging to this category. In his article, Falk Müller shows how two large German companies, Siemens & Halske and AEG, became involved in a struggle to dominate the field of electron microscopy. Siemens in particular was prepared to give its researchers considerable support to win this battle of prestige, which included quarrels about patents, promotional strategies and marketing campaigns – as if both companies were competing in the context of a free market economy rather than under a totalitarian regime in wartime.

With varying degrees of success, the companies tried to convince the authorities of the prestige the electron microscope lent to the German sciences and its relevance for the war effort, not only to promote the instrument, but also to seek support and to prevent researchers from being drafted into military service.

The second group are the Dutch geneticists and eugenicists as portrayed by Stephen Snelders. He sets out that both 'respectable' and Nazi geneticists and eugenicists often endorsed the same principles and aims of racial hygiene. Forced sterilisation and medical examinations before marriage were topics discussed by geneticists and eugenicists from all parts of the political spectrum. Only in the course of the war did the distance widen between Nazis and non-Nazis, but this reflected the increasing alienation between the German occupier and the Dutch population rather than the contents of the eugenic ideas.

Certainly, a difference between non-Nazi and Nazi eugenicists and geneticists was the latter's ideological motivation. The Nazis worked with the urgent feeling 'that survival of their race was at stake'. Indeed, as Mark Walker argues in his article, the Germans fought a racial war and considered anthropology and genetics of vital importance to overcome – no less than arms. The Dutch non-Nazi eugenicists, however, who did not feel this urgency, remained somewhere in the twilight zone between the scientists working for the war effort and those who were marginalised.

Science under occupation

As the reader will note, the location of the conference has a heavy emphasis in the composition of this book: a large part of it is devoted to 'Dutch' people and locations. The interesting thing about it is that so far little attention has been paid in the international historiography to science that was conducted in occupied territories. The totality of 'Dutch' contributions might, therefore, be considered as a case study of science in an occupied country.

After the conquest, the Nazis introduced a civil administration in The Netherlands. Considering the Dutch as a related (Germanic) race, they hoped to win over the Dutch population to their cause and ideals. They therefore adopted a reserved attitude. The Dutch initially reacted compliantly towards their occupiers. They adapted themselves to the situation that had arisen – thereby neither actively opposing the occupiers, nor embracing their cause – and aimed to continue their lives in an undisturbed manner. It was only after the war prospects worsened for the Germans, who increasingly exploited The Netherlands for their war effort, that the Dutch become more and more anti-German. In the final phase of the war (from summer 1944 on) the occupation degenerated into terror and plundering and a complete disorganisation of Dutch society (The Netherlands, moreover, became a front-line state).

The articles on Dutch scientists and institutes in this collection reflect these circumstances and changing attitudes. The Dutch scientists adapted to the new regime in a compliant manner and tried to continue to work without interruption. Increasingly, there was a threat of their instruments and workers being taken away by the Germans and, increasingly, the scientific institutes allowed illegal activities within their walls.

These traits of the Dutch situation bear important similarities with that in France. As in The Netherlands, scientific life apparently continued undisturbed, despite shortages and broken contacts. In France, too, the attitude on the part of the scientist changed from compliant to increasingly hostile.[8] However, it is clear that in other occupied areas the situation could be different. In the Eastern European countries in particular, with their in German eyes 'inferior' Slavic populations, the occupiers ruthlessly annexed, plundered and destroyed scientific institutions and resources without reservation and left the scientists with no other choice but to work in their interest.[9]

Similarities

Apart from the very different situations scientists could find themselves in, their behaviour had remarkable similarities. These concern their moral behaviour, for example. The international comparative studies in the collection *Politics and Science in Wartime* (2005) have provoked its editors, Mark Walker and Carola Sachse, to sketch a not very lofty ethical image of the wartime scientist. Scientists everywhere were in the first place driven by self-interest. This was expressed in the opportunistic way they accommodated themselves to the political framework to retain optimum support for their research and position. The 'scientists exploit as far as they can the particular professional opportunities offered them by the state, without considering the possible political or military consequences and without any ethical scruples'.[10] This book, which with its 'down-to-earth' case studies employs a 'bottom-up' approach, only underlines the amoral image of the sciences in wartime set out by Walker and Sachse's 'top-down' study.

Indeed, the most appropriate key words to explain the behaviour of the majority of the scientists in this volume are self-interest, opportunism and accommodation. With the exception of Occhialini, all the scientists – from Wander Johannes de Haas to Yoshio Nishina – who were confronted with a regime that interfered with their working conditions made a similar choice. Whether they supported the regime or not, they accepted the rules imposed by it, and even repression, in order to protect their position, lab and employers, and even took profit from the situation to optimise their research circumstances. Resistance activities were only organised in the Dutch laboratories when the Germans were about to lose the war.

As Walker sets out in this volume, the traditional inclination to judge the scientists' behaviour by labelling it 'good' or 'bad' does not satisfactorily help to interpret their deeds. De Haas, on the one hand, urged his students and employees to refrain from action against the occupiers and even worked for the Germans. But, on the other hand, the maintenance of his laboratory created the opportunity to organise illegal activities there. Similarly, the fact that the Dutch aerodynamic research institute NLL could survive by supporting the German war effort enabled it not only to help its staff through the direst years of the occupation, but paradoxically also to become a place where resistance activities (such as the production of arms) were organised. The dichotomous good–bad approach cannot be applied to interpret these situations satisfactorily – self-interest, opportunism and accommodation can.

Another recurring observation is that of the previously mentioned continuation of much of the research, especially in the early years of the war.[11] This was not only an expression of the general inclination to hold on to daily routines, but also often to keep things going to prevent students and staff from being drafted into military service or forced labour. It sometimes resulted in over-employment in the laboratories, which could even lead to an increase in research activity. This way of protecting lab and staff occurred in both occupied and war-waging countries. The best thing to do to continue research was to convince the authorities of its importance for the war effort.

Continuity, however, also reflects the ingenuity and flexibility to cope with difficult circumstances. The case of Delft penicillin shows that isolated scientists were still able to acquire (though in this case not without luck) the necessary information. The workers at the Kamerlingh Onnes Laboratory managed to keep liquid helium available until well into 1944. And Nishina, who depended heavily on American know-how, nevertheless ultimately managed to make his cyclotron work.

Next, while the researchers working for the war effort were often highly motivated, as mentioned above, this could also be the case for marginalised scientists in occupied countries. The humiliating occupation often gave rise to a strong nationalistic sentiment in those societies. In this atmosphere the scientist could be determined to preserve national pride and spirit by making achievements in research or by teaching future generations, rather than being discouraged by being cut off from contacts, recent literature and materials. In France, the occupation led to an 'intensified sense of moral duty' to maintain a 'normal scientific life'.[12] In Italy, as Gariboldi writes in this book, unpopular measures by the regime provoked in the scientists a 'hidden revolution' to 're-conquer' the 'Italian soul'.[13] The scientist might not have excelled by his very principled behaviour; the events of the war did, of course, not leave him untouched.

Finally, scientists who were cut off from their daily routine often broke new ground in an unexpected manner; improvisation sometimes resulted in surprising new findings. Le Poole maintained that he would not have succeeded if he had known about the latest developments in electron microscopy in the United States and Great Britain. Here, the isolation was splendid indeed. The development of Bacinol would never have been started if a lack of material to fill their fermenters had not forced the NG&SF to look for new opportunities. Yet, such a spirit of innovation induced by extraordinary circumstances was not unique to occupied countries. Scientists introduced to war-related projects could become inspired by their new environment and working with people from different backgrounds. As set out by Yamazaki in this volume, Shinichiro Tomonaga thought up his general theory of ultra-short-wave circuits while working with the engineers of the Japanese Navy Technical Research Institute. Peter Galison has even contended that the famous Feynman diagrams were inspired by the influence of the Los Alamos environment on the way of thinking of Richard Feynman.[14] The cryptographic work at Bletchley Park may have stimulated the pioneering work on the computer in Manchester shortly after the war (in which former Bletchley Park workers were involved), though Field argues in her article that it is impossible to make an unequivocal judgement about this.

An opportunistic mind was apparently a prerequisite for a successful wartime scientist, not only to survive the political situation, but also to cope with the difficulties caused by the war and to break new ground. The fact that scientists, ending up in such different situations, nevertheless behaved so similarly in some respects, might on closer inspection not be that surprising. After all, they had one important thing in common: they were all 'ordinary' scientists in extraordinary circumstances.

Structure of the book

After Mark Walker has set the stage with his survey in Chapter 1, we continue this collection with, in our view, the most marginalised scientists: Giuseppe Occhialini as described by Gariboldi (Chapter 2), followed by Burns' isolated Delft penicillin team (Chapter 3) and Van Delft's Kamerlingh Onnes Laboratory physicists (Chapter 4). We will then gradually proceed to less marginalised examples: Fournier's Dutch electron microscope builders, who at least had contacts with German colleagues, (Chapter 5) and, in Chapter 6, the aviation physicians of von Lünen (a difficult case, concerning both the very marginalised Jongbloed and the far less marginal German community of aviation physicians). Snelders' eugenicists (Chapter 7) and Müller's German electron microscope researchers (Chapter 8) were neither marginal nor working strenuously for the war effort. Florian Schmaltz's scientists of the NLL (Chapter 9) are the first scientists we encounter actually working for the war effort. This paper is followed by the articles about Nishina's student, Takeuchi, by Yamazaki (Chapter 10) and Nishina himself by Nagase-Reimer (Chapter 11). Of particular importance for the course of the war, and therefore concluding this book, are Bissell's article about feedback control (Chapter 12) and Field's study of Bletchley Park (Chapter 13).

Notes

1 For a historiographic survey, see Mark Walker's article in this volume.
2 R. M. Macleod (ed.), *Science and the Pacific War. Science and Survival in the Pacific, 1939–1945* (Dordrecht, Boston, London: Kluwer Academic Publishers, 2000); M. Szöllösi-Janze (ed.), *Science in the Third Reich* (Oxford, New York: Berg, 2001); U. Deichmann, *Biologen unter Hitler: Porträt einer Wissenschaft im NS-staat* (Frankfurt am Main: Fischer Tachenbuch, 1995); D. Edgerton, *Warfare State: Britain, 1920–1970* (Cambridge: Cambridge University Press, 2006). An account of the publications of the Kaiser Wilhelm Society project: http://www.mpiwg-berlin.mpg.de/KWG/publications.htm (last accessed 10 July 2007).
3 K. Bird and M. J. Sherwin, *American Prometheus: The Triumph and Tragedy of J. Robert Oppenheimer* (New York: Alfred A. Knopf, 2006), 279–81.

4 Even in the 'sharashkas' in the USSR, where imprisoned scientists were exploited as convicts, successful research was conducted: E. I. Kolchinsky, 'Science Mobilization in the Soviet Union', *Historia Scientarum* 16 (2006), 15–28, 17 and 25.
5 W. E. Grunden, Y. Kawamura, E. Kolchinski, H. Maier and M. Yamazaki, 'Laying the Foundation for Wartime Research: A Comparative Overview Mobilization in National Socialist Germany, Japan, and the Soviet Union', C. Sachse and M. Walker (eds), *Politics and Science in Wartime. Comparative International Perspective on the Kaiser Wilhelm Institute* (Chicago: University of Chicago Press, 2005), 93–6; W. E. Grunden, M. Walker and M. Yamazaki, 'Wartime Nuclear Weapons Research in Germany and Japan', Sachse and Walker 2005, 107–30.
6 As is argued for the French case in: N. Chevassus-au-Louis, *Savants sous l'occupation. Enquête sur la vie scientifique francaise entre 1940 et 1944* (Paris: Seuil, 2004).
7 Kolchinsky 2006, 25.
8 Chevassus-au-Louis 2004, 21–81.
9 O. Elina, S. Heim and N. Roll-Hansen, 'Plant Breeding on the Front: Imperialism, War and Exploitation', Sachse and Walker 2005, 161–79.
10 C. Sachse and M. Walker, 'Introduction: A Comparative Perspective', Sachse and Walker 2005, 1–20.
11 This continuity has also been observed for science in France: Chevassus-au-Louis 2004, 27–30.
12 Chevassus-au-Louis 2004, 29–30; for The Netherlands, see for example: L. Molenaar, *Marcel Minnaert, astrofysicus 1893–1970: De rok van het universum* (Amsterdam: Balans, Leuven: Halewyck, 2003), 280–1.
13 A comparable attitude can also be found at German non-Nazi scientists, see for instance: J. Medawar and D. Pyke, *Hitler's Gift: Scientists Who Fled Nazi Germany* (London: Richard Cohen Books, 2000), 157–89.
14 P. Galison, 'Feynman's War: Modelling Weapons, Modelling Nature', *Studies in History and Philosophy of Modern Physics* 29 (1998), 391–434.

1 The mobilisation of science and science-based technology during the Second World War

A comparative history

Mark Walker

Science during the Second World War is a large topic that has received a lot of attention from historians, but the coverage has been uneven.[1] While some subjects have been studied in detail, others have barely been touched, and there is still much to discover or reinterpret. This essay will try to use as broad a brush as possible in order to set the stage for other chapters in this book as well as to give the reader a sense of the scope of the subject.

My overview will emphasise certain countries and developments in science and technology over others, and does not claim to be a definitive account. As Roy Macleod has noted in the introduction to a collection of essays on science and World War II in the Pacific theatre:

> many Western historians of science see the emerging relationship between science and war as a ... regrettable interruption in the normal flow of scientific discovery, rather than as an expected 'norm' of practice. In this perspective, such relationships are seen as unwarranted intrusions – their existence revealing a dark, unnatural side of the Enlightenment project that viewed reason and science as agencies of universal improvement ... military connections were a necessary evil, in the long run detrimental to the ethics and interests of science, and possibly of civilisation itself ... We are now encouraged to read the history of science and war as parallel, often intersecting, and mutually dependent activities, rather than opposing narratives.[2]

In a similar vein, David Edgerton has recently published a revisionist account of science in Britain during the Second World War, arguing that Britain was a 'warfare state' between 1920 and 1970, and that historians and scientists alike have overlooked a great deal of the science because it was not located in the usual academic places: '... important state and industrial military laboratories, design centers and workshops were responsible for most of the major [British] innovations in military technology of the Second World War'.[3]

The literature on the other winners of the Second World War is not satisfying. There are several good recent works on Soviet science, but with surprisingly little coverage of science in service of the Second World War.[4] The Soviet atomic bomb is, of course, a post-war development,[5] and most literature on Soviet science is arranged thematically such that the war does not stand out. Yet science and science-based technology must have made an important contribution to the eventual Soviet victory over Germany on the eastern front.

As far as the United States is concerned, older literature such as Daniel Kevles' account[6] still dominates the historiography. In general, there is an emphasis on such 'Big Science' projects such as the atomic bomb[7] or radar.[8] However, the tacit assumption that these two novel, interdisciplinary, and large-scale projects were typical or even stereotypical of American science – let alone science in general – in the service of the Second World War should now be questioned. This assumption has been strengthened by the recent emphasis placed by historians on science in America after World War II.[9] This emphasis may well be due to the intrinsic importance of the topic, but an assumption that we already know what there was to know about the wartime period may also have contributed.

Recently the losers, Germany and Japan, have attracted the attention of historians of science interested in the war.[10] In a book on Japanese secret weapons, Walter Grunden argues that:

> Japan's failure to organise large-scale research and development projects for the war effort stalled its progress toward ... high-technology development. Although Japan also engaged in efforts to produce nuclear weapons, radar, and technologically sophisticated missiles, it did not effectively mobilise its pool of scientific and engineering talent en masse towards the end ... there was no Big Science revolution in Japan as occurred elsewhere in the West during World War II.[11]

The German case is relatively well known. There is an older but still valuable book from the 1970s on the mobilisation of engineering and technology.[12] During the 1980s and 1990s, literature appeared on the German variant of 'Big Science', including aeronautics, rocketry, and nuclear research.[13] Disciplinary studies appeared at around the same time for individual sciences, although usually not emphasising the theme of World War II.[14] The Max Planck Society's research program for the history of (its predecessor) the Kaiser Wilhelm Society during National Socialism produced a great deal of work on German science during the Second World War – the war made up almost half of the Third Reich – as well as a major work by Helmut Maier on the mobilisation of science and science-based technology for the war effort.[15] The important message of this newer literature is that rather than the war stopping or interrupting science in Germany, it often enhanced and facilitated it.

Unfortunately, very little is known about the application of science to war for Axis ally Fascist Italy,[16] wartime allies such as Hungary, or collaborating regimes such as Vichy France.[17] Émigrés played an important role in the fight against Germany, but this subject is usually integrated into histories of particular projects, rather than studies of émigrés in particular, so that it is also mainly limited to subjects such as radar or atomic bombs.[18]

Recently scholars have used comparative history in order to investigate the interaction of science and war.[19] For various reasons German examples have played a major role in these comparisons, with a welcome integration of Japanese case studies. Just as with national histories, there have been relatively modest contributions from experts on Soviet or American science.

In general, the subject of science during and contributing to the Second World War is surprisingly incomplete. The fact that some subjects such as radar, rockets and nuclear weapons have received so much attention obscures the fact that a great deal is either poorly understood or essentially unknown. This is both the result of, and the reason for, the common assumptions that radar and the atomic bomb are all one needs to know, that they are the patterns the rest of science followed. This collection serves the valuable purpose of fleshing out and expanding this picture.

Scope of Subject

What should be included and excluded from this study? Scientific research or development work that was no different from research done in peacetime is relevant in this regard. Sometimes, a remarkable amount of continuity in the research existed where one might have expected an interruption because of the extraordinary circumstances created by the war. The historian of science is interested in both this 'unexpected' continuity and the way scientists managed to maintain it.[20] Research done because of the war, directly or indirectly, or done significantly differently because of the war, is especially important. This includes research on weapons and other things directly relevant to the war effort, but should not be restricted to it.

Time span is also important. Studies of science during World War II cannot merely begin with the commencement of hostilities and end with the armistices – and not only because, depending on the country, there is no one beginning or end. There is a relevant prehistory, as well as postscript to this history. The former should include, for example, rearmament and mobilisation efforts immediately prior to the war; the latter should also examine the post-war transfer of technology, scientific techniques, and manpower from Germany to other countries. But, of course, the line has to be drawn somewhere. This chapter will include the Soviet work on atomic bombs from 1945 to 1949 because this was an immediate consequence of wartime developments, but other post-war examples will be excluded.

Who should be included is also an important consideration. Of course, scientists from the warring countries, but also scientists in occupied countries as well as those scientists forced to work for the German war effort, both inside and outside of Germany. Ideally, scientists working in industry in all of these countries should be included. They contributed an important and large portion of the research and development during the war, although much less is known about these researchers because of the difficulty in gaining access to sources in private archives.

Of course, historians would want to include research that today is, and perhaps even immediately after the war, was recognised as important, first-rate science. But they should also look at work that scientists today might not be impressed by, that might now even be labelled 'not serious' or 'pseudo-science', so long as the researchers at the time did seriously pursue it, or the regimes seriously supported it. Thus, in general, a history of science during World War II should include much more than it excludes.

The Quality of the War

World War II was not the same everywhere or for every side; rather, it was different for the various countries involved. In Germany it was also a racial war. Even if we limit ourselves to the period after the German invasion of Poland, German military strategies and occupation policies clearly had a racial and sometimes genocidal dimension. Scientists were intimately involved, for example, in the planning and implementation of the 'Germanisation' of occupied lands in the east. Scientists helped sift the 'racially valuable' element out from the existing populations and plan the future settlements for transplanted Germans. When German anthropologists trained SS physicians in racial hygiene (the German term for eugenics) during the war, this was as direct a contribution to the war effort as German engineers training soldiers how to operate a new weapon. Scientists who were trying to develop a scientific method for determining race, given the context of a war against racial enemies, inside and outside of the National Socialist 'People's Community', were also not only contributing to the war effort, but were seen as doing precisely that.

In the Soviet Union, ironically, it was not an ideological war of communism against fascism. Early on, Stalin and other leading Soviet officials realised that this would not suffice to rally the Soviet people. Instead, the Soviet leadership called upon their citizens to fight for mother Russia against the invading Germans – and here, as in most countries fighting Germany during the war, 'Nazi' and 'German' came almost to be synonyms. In Japan it was a racial war, including the brutal occupation of China and other parts of East Asia, as well as a fight to ensure Japan's rightful position as a world power. In the United States, the war was not about German racism or anti-Semitism, rather German aggression, but was racist towards the Japanese. In Britain it was not a racial or ideological war, rather more a matter of responding to German aggression. Since the war meant different things to different countries, it also meant different things to scientists in these countries.

Total War

In total war, when whole societies or populations are mobilised for the needs of a modern, industrial and technological conflict, almost all scientific activity becomes war work. During wartime, strained war economies both lost manpower to the army and were urged to increase the quality and quantity of production. In the warring countries, Germany, Japan, the Soviet Union, the United Kingdom and the United States, these mobilisations led to a shortage of scientists and engineers for industries vital to the war effort.

The universities and technical colleges were expected to help make up the difference, even as their staff and students were being called up. Where possible, other scientific manpower was mobilised and brought to the universities, and other types of students, sometimes including more women, were enrolled. Thus, scientists who 'merely' taught beginning chemistry or physics to undergraduates were making an important contribution to the war effort, and this was recognised as such. The German case provides two good examples of this.

When a Nazi colleague, Johannes Stark, denounced the theoretical physicist Heisenberg in the mid-1930s as a 'White Jew' and 'Jewish in Spirit', his case was investigated by the SS. In the spring of 1939 Heinrich Himmler himself informed Heisenberg that he was being rehabilitated. A separate letter from Himmler to another SS official made clear that this was not because of Heisenberg's contributions to modern physics, but rather because of his value as a physics professor: he had already trained a school of talented young physicists, and if he stayed in Germany he would continue to do so.[21]

During the Third Reich, the German Physical Society gradually lost most of its autonomy. When the tide of the war began to turn against Germany, the society offered to help mobilise physics for the war effort and insisted that the German state in turn needed to provide greater support for physics. This mobilisation of physics emphasised both the potential of physics to help create novel weapons and the vital role the teaching of physics played for the training of scientists and engineers for industry.[22]

The Mobilisation of Science for War

Recently, scholars have compared the mobilisation of science for the war in Japan, Germany, the United States and the Soviet Union. Perhaps the most important step they took was their decision not to take the United States, or the Manhattan Project and the Rad Lab as the standard of comparison. The different warring countries had very different political systems, ranging from Fascism on the right and Communism on the left, with liberal democracy and an authoritarian Empire somewhere in the middle. Similarly, both the economies and science were organised differently as well.

Yet when one compares the mobilisation of science in Germany, Japan, the Soviet Union, and the United States, an interesting pattern emerges. In each case, the need for interdisciplinary mobilisation of science for the war was solved by creating inter-institutional committees staffed by technical experts, not political appointees, to solve particular problems that arose in the course of research and development. There were, of course, real rivalries and inefficiencies as a result, but these have arguably been emphasised more than the examples of harmonious cooperation.[23]

Even in the Soviet Union, where scientific research was already centralised, for example at the Soviet Academy of Science, such committees had to be created in order to make the research more efficient and increase production. In Germany, famous for its apparently chaotic war economy, we find that, while the ministers were fighting with each other, at a level below their technically and scientifically trained subordinates were cooperating. In the United States, the engineer and political advisor Vannevar Bush created such committees through the Office of Scientific Research and Development to help him reorganise American science for the war, and Japan did as well, if on a smaller scale. Of course, not all topics or fields were organised successfully in all countries, but it is striking how much was done relatively well.

The word 'relatively' points to a fundamental problem: should we only judge things according to the standard of the victor? In particular, the United States was alone among the warring countries to have ample manpower, resources, and, perhaps most important, after Pearl Harbor was not attacked on the home front. Moreover, it had the largest economy to begin with. In the other countries, one goal often came into direct conflict with another. Thus, in Germany, the mass conscription of young men as soldiers conflicted with the mobilisation of scientists. Eventually, this led to a belated recognition that the conscription of scientists and engineers had gone too far, and a recall of scientists and engineers from the front was made in 1943–1944. These recalls were only possible for projects considered important for the war effort. For example, Peter Wegener was brought back in order to work on wind tunnels at Peenemünde and elsewhere.[24]

Universities

University life and science, both as research and instruction, was severely disrupted by the mass mobilisation of young men, or the closing or restriction of universities in occupied countries. However, this also has to be differentiated. The war meant different things to scientists at different points in their careers. Senior, established scientists could more easily stay at their positions, and some could even avoid war-related research. Junior scientists were more at risk of being called up and had to avoid direct military service. The best way to do this was usually to get involved in a research project supported by the military or industry under contract from the military. Science students often were called up before they had finished their degrees, so that they had little chance of getting an exemption from direct military service. The consequences of this could be seen very clearly in Germany after the war. The post-war German universities were flooded by students who were usually either very young or rather old – the generation them between had been lost.

For the researchers who did remain at the universities or other state-run institutes, it was often feast or famine. The scientists who were able to stick with pre-war research that was not considered important, or even relevant for the war effort, often found that their research assistants, materials, and experimental apparatus dried up – these are precisely the scientists who are not well known for their wartime work. The aeronautical expert Ludwig Prandtl, on the other hand, whose research was directly relevant for the development of new military aircraft, found that when it came to support from the National Socialist government, he could not ask for too much.

'Big Science'

When we think of changes in scientific practice during World War II, one of the first things that come to mind is the great scale of some of these scientific projects, including the research and development of ballistic rockets at Peenemünde in Germany, of radar devices at the Rad Lab in Cambridge, Massachusetts, and of atomic bombs in Los Alamos. It is true that historians have argued that at least some of these were not merely created by the war, rather were the end products of long-standing trends in this direction, but there nevertheless is a profound difference in the scale, and sometimes quality of scientific research during the Second World War when compared to what came before it.

Scientists joining the Rad Lab or coming to Peenemünde entered another world, where exciting and innovative interdisciplinary research was possible on a large scale with talented colleagues and generous support. Very often, this was 'technically sweet' work, in the words of Robert Oppenheimer, research and development that was interesting and fulfilling. There are many examples, from all of these projects in the different warring countries, of scientists and engineers who were able to shut out both the war and the military significance of what they were doing and instead throw themselves into their work. They were aware that they were doing something important, perhaps unprecedented, but did not face all of the consequences.

'Ideologically Correct Science'

'Ideologically correct science', as the name suggests, was analogous to what today is called 'politically correct' behaviour, although the stakes were higher. In three different countries involved in the Second World War, a similar pattern emerges. During the 1930s in Germany, National Socialist scientist-activists denounced what they called 'Jewish' science and called for an 'Aryan' science. In the Soviet Union during the 1920s and 1930s, Communist scientist-activists denounced 'bourgeois' science and the so-called 'bourgeois specialists' and called for a 'Marxist' science. The Soviet case culminated in the so-called 'Lysenko Affair', whereby biologists and agricultural scientists advocating modern genetics were purged. During the 1930s in Japan there were calls, although usually not from scientists, for a 'Japanese' type of technology that was not derived from Western science.[25]

In all of these cases, the ideologues enjoyed initial success, but the beleaguered 'orthodox' scientific communities managed to beat back these attacks and eventually silence them by finding their own patrons within the political leadership of their country and, in particular, demonstrating their usefulness to the goals of that leadership. Perhaps the best example of this is provided by the nuclear weapons projects in Germany, Japan and the Soviet Union.

In general, the need for scientists during the war provided a niche for some scientists who were out of favour with their regimes. In Germany, some scientists who were banned from academic jobs for political or racial grounds got jobs in industry or as independent contractors for industry doing war-related work.[26] In particular, the status of an independent contractor appears to have allowed research institutions and industrial firms to work with scientists they could not otherwise have employed.

In the Soviet Union, efforts among physicists to emulate Lysenko and his followers in biology and agricultural science and highjack their discipline were stopped because the regime recognised that it needed the physicists and modern physics for the atomic bomb project.[27] Military officials in Japan overcame their mistrust of Western-trained scientists in order to begin work on nuclear weapons. In the United States, the unorthodox physicist Leo Szilard was allowed to contribute to the project even as Groves and the American FBI kept a close watch and disapproving eye on him. General Groves also picked Oppenheimer to be the Scientific Director of the new Los Alamos Lab despite knowing about his connections with communists.

This assessment of Oppenheimer's case is an inversion of the usual interpretation of the relationship between Groves and Oppenheimer.[28] The orthodox account accepts without question that Groves selected Oppenheimer during the war to run the Los Alamos weapons lab, and portrays Groves' subsequent lack of support for Oppenheimer during the 1950s when his security clearance is taken away as a betrayal. One can instead interpret this as Groves having had misgivings about Oppenheimer's connection to communists from the beginning, setting them aside during the war because he needed him, and then returning to them when the American government questioned Oppenheimer's loyalty in the 1950s.

In an earlier work on the German nuclear weapons project during the war, this author argued that scientists would serve any master.[29] When the broader context of science policy and administration in the different nations at war is taken into account, it also appears that, especially during war time, these governments or masters in turn were willing to employ those scientists willing and able to serve them.

Science without Moral Boundaries

In some countries, total war also relaxed the moral boundaries that usually restricted scientific research, if it might further the cause of the war. Japanese researchers developed and tested biological weapons on prisoners in occupied China.[30] The German air force sponsored experiments on concentration camp prisoners.[31] In the latter case, it was not so much the type of experiments that was unusual – the Americans and Canadians carried out similar experiments. However, there was a difference in the German case, for the experiments were deliberately taken to the point of killing the subjects. Other researchers, including Josef Mengele and August Hirt, experimented on Jews, Soviet prisoners of war, and others in the camps.[32] The difference in scientific practice here is the availability of large numbers of human subjects who could be and were treated as disposable.[33]

Biological weapons and concentration camp experiments are the best-known examples, but scientists used people without their consent as research objects in other cases as well. The so-called 'Euthanasia Campaign' in Germany during the first few years of the war was carried out first and foremost in order to free up other resources and was therefore directly related to the war. The geneticist Hans Nachtsheim 'borrowed' children slated to be killed for some experiments on artificially induced epilepsy. Even if the experiments themselves did not harm the children, they also did not participate voluntarily.[34] Here it is not merely a matter of moral boundaries being transgressed in the course of research directly relevant to the war effort, but also of research that was facilitated by the war.

'Wonder Weapons'

Scientists in all warring countries worked to help develop weapons or materials to help their side win the war (or not lose it). But some projects were more fantastic, and aimed perhaps to win the war outright. The Germans called these 'Wonder Weapons' and, echoing Propaganda Minister Josef Goebbels' propaganda, sought to overcome their quantitative inferiority through qualitative superiority.

The most effective of these weapons were radar and atomic bombs, achieved first by the Americans with the help of British and German émigrés. The Soviets did not turn to such weapons until after the war, simply because up until that time they needed their scientists and engineers to help develop and produce conventional weapons (which were enough to defeat Germany). Both the Japanese and

Figure 1.1 The Kaiser Wilhelm Institute for Chemistry, after an allied bombing raid.

the Germans pursued these weapons in desperate last efforts to turn the tide of the war. The Japanese versions of such weapons included suicide weapons but, of all the ones investigated, few were actually finished.

It was in Germany that the search for wonder weapons was the most desperate. This included rockets,[35] jet planes[36] and nuclear weapons.[37] But it went further than that. One of the most interesting aspects of German science during the last phase of the war is that in almost every discipline or subdiscipline one can find groups of scientists and engineers who struggled to use whatever expertise and materials were at hand in order to create some sort of weapon to somehow help influence the outcome of the war. This included improbable weapons such as the particle beam gun that would be used to shoot down enemy planes,[38] but also the nerve gases that German physical chemists invented and which still plague us.[39]

By 1944, most of the leading researchers in the German nuclear weapons project assumed that atomic bombs could not be built before the end of the war. (Ironically, they also took for granted that they were ahead of the Americans in this regard.) They had not slowed down or diverted their research, but they also were not pushing it as hard as they could have. However, a small subset of this project, working in secret under the leadership of the Army Ordnance physicist Kurt Diebner, tried to design a weapon – not an atomic bomb – using conventional high explosives as well as nuclear fission and fusion reactions. Shortly before the end of the war, these researchers apparently tested such a device. Although it is not clear that this weapon test was successful, in other words, that it produced fission and fusion reactions, arguably what is most significant about this episode is that they even tried.[40]

Collaboration

The recent controversy surrounding Peter Debye in The Netherlands clearly shows that the topic of collaboration by scientists with the enemy is a sensitive, yet also important, subject.[41] First of all, the term 'collaboration' should not be restricted to scientists we do not like, or science that is applied or military.[42] Academic

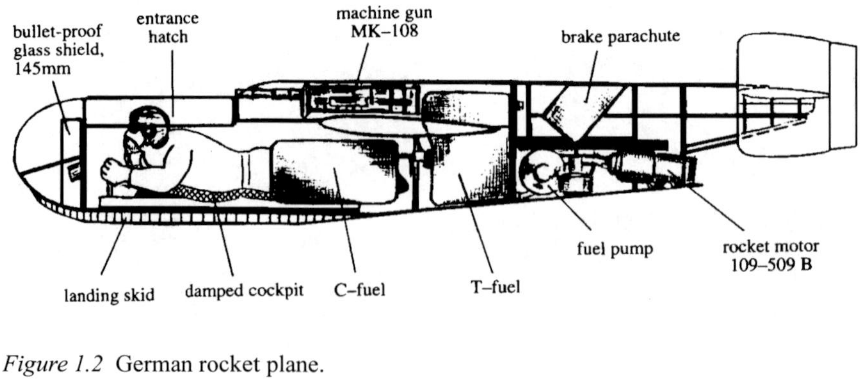

Figure 1.2 German rocket plane.

researchers were under direct pressure to collaborate, simply because the occupying governments were particularly concerned with controlling the universities where they worked. Their work could be quickly and effectively stopped by closing down their labs and institutes. When assessing the extent to which scientists in occupied countries collaborated with the occupying power, this should not be limited to academic scientists or students, for in every country many scientists were working in industry. Most industrial firms in occupied countries apparently collaborated with the German authorities, so one can reasonably assume that their scientists did so as well.[43]

As the papers in this collection show, there were clear-cut examples of collaboration, as well as equally clear examples of scientists who refused, openly or secretly, to do so. At the conference in January of 2007 at the Museum Boerhaave in Leiden, some Dutch and French participants argued that, from their perspective, collaboration was either essentially 'good' (justifiable or acceptable) collaboration or 'bad' (unjustifiable or unacceptable).

This struck a cord with this author, for it reminded him of the discourse surrounding German science under Hitler that he had encountered when he lived in West Germany in the 1980s. Here, the issue was not 'collaboration', for one does not usually describe German scientists working in Germany as 'collaborating' with the National Socialist regime, rather of being either for or against the Nazis.

Indeed this goes further back in the literature on science under Hitler. In August of 1945, Samuel Goudsmit, naturalised American physicist and scientific head of the 'Alsos Mission', a scientific intelligence-gathering mission sent out to find and liquidate the threat of a German atomic bomb, described the German scientists he encountered as follows:

> Otto Hahn ... has been very strongly anti-Nazi ... [Paul] Harteck is a fine scientist of high character and is strongly anti-Nazi ... [Carl Friedrich] von Weizsäcker ... is anti-Nazi but makes compromises fairly easily ... [Werner] Maurer is generally conceded to be a disagreeable, extreme Nazi ... [Rudolf] Fleischmann is a Nazi and not very well liked by his colleagues.[44]

The émigré physicist James Franck described Hahn in similar terms: 'I inquired of many places about my old friend, Otto Hahn, who, as you probably know, was always an anti-Nazi ...'[45]

When it came to describing 'Nazi' colleagues, Goudsmit was not always kind: '... a first-rate windbag and second-rate physicist named [Ludwig] Wesch ...This man was the top Nazi in Heidelberg...'[46] Similarly, according to Goudsmit the respected technical physicist Abraham Esau '... had attained his post mainly by being an ardent Nazi'.[47] When the theoretical physicist Arnold Sommerfeld retired in the 1930s, he was: '... replaced by a Nazi named [Wilhelm] Müller, who did not "believe" in modern physics ...'[48]

In contrast, Goudsmit had some good things to say about some of his German colleagues. '[Wolfgang] Bothe was a loyal German but never a Nazi'.[49] 'Von Weizsäcker ... was not a real Nazi, but like his father was a real diplomat. He knew how to strike a compromise with the Nazis whenever it was expedient'.[50] '[Walter Gerlach] was not a Nazi, but on many occasions, we learned, his judgment had failed him and in order to avoid trouble he had played into the hands of the Nazis ...'[51] When it came to Goudsmit's respected colleague Werner Heisenberg, Goudsmit noted that he had worried whether: '[Heisenberg's] contact with Nazism had not changed him too much. He had already lost the confidence of several of his anti-Nazi colleagues'.[52] Some Germans, like Hahn, even earned the title of 'anti-Nazi' from Goudsmit: '[Wolfgang] Gentner himself was an anti-Nazi who had worked in the United States with Ernest Lawrence ...'[53] 'During the entire Hitler regime, including the war years, [Max] von Laue had openly opposed the Nazis in his actions and utterances'.[54]

It is perhaps not surprising that Goudsmit, writing immediately after the end of the war and the revelation of the Holocaust (he lost his own parents in Auschwitz), saw his colleagues in a polarised picture of black and white, of 'Nazis' and 'anti-Nazis'. However, this dichotomous model remained in place for decades thereafter. In an excellent and insightful 1969 book by the political scientist Joseph Haberer the same terminology is applied.

> More than any other institutional or professional community in Germany, the scientists disengaged themselves from the problem of their responsibility in the crisis which involved them ... Non-National Socialist scientists who cooperated with the regime and were obviously implicated in some of its policies, took a slightly different position: we did what had to be done to save German science, which was our duty to our profession and to our country ... Flight as a form of protest was rejected, since to leave Germany would be to leave its future to those least qualified to provide for it – National Socialist scientists like Stark and [Philipp] Lenard.[55]

> Those scientists who were anti-Nazi or non-Nazi invariably assumed that any open resistance would be useless and self-defeating, and that only continued cooperation with the regime, even under the most odious conditions, would make it possible to avoid the total destruction of German science ...[56]

> The real issue involves how it was possible for men trained in the sciences, like Lenard and Stark, to become fanatical National Socialists.[57]

During the 1970s and 1980s, historians working on German scientists 'under Hitler', in Alan Beyerchen's phrase, slowly pulled themselves away from the dichotomy of black and white, good and bad, and eventually developed the approach that is standard, if not required today: rather than judging whether an individual or case was 'Nazi' or 'anti-Nazi' (or good collaboration or bad collaboration), the individual or case is examined in both detail and context so that in the end one does not say an individual was a 'Nazi' or not, rather one describes and analyzes what this individual did and why.[58]

In this spirit let us take an admittedly ambivalent example of collaboration between the two physicists Werner Heisenberg and Hendrik Kramers.[59] In the fall of 1942, Heisenberg had made a controversial trip to occupied Denmark that clearly alienated his Danish colleagues. Thus, when the collaborating Dutch Ministry of Culture contacted him in 1943 and invited him to give lectures in occupied Holland, he reacted cautiously. Who was inviting him, and what did his Dutch colleagues think about it? Heisenberg apparently did not want to make the same mistake twice.

The response came directly from Kramers in a private letter: Heisenberg's Dutch colleagues wanted him to come. It would not be a matter of public propaganda; rather, Heisenberg would visit physics institutes and give talks for his fellow physicists. Moreover, Kramers added, the Dutch physicists hoped that Heisenberg's visit could help improve their situation. Many of them were locked out of their labs and institutes because of their opposition to some occupation policies.

Heisenberg came, and did just what Kramers had proposed. In the end, the Dutch physicists were ambivalent about Heisenberg's visit. On the one hand, he clearly was trying to help them by doing what they had suggested. On the other hand, while in Holland, and in situations where it had not been necessary to use pro-Nazi or pro-German rhetoric, he told Hendrik Casimir that it was a choice between (National Socialist) Germany and (Communist) Russia, and 'perhaps an Europe under German leadership would be the lesser evil'.[60]

Figure 1.3 Werner Heisenberg, in the middle, on a training exercise before the war.

Shortly after Heisenberg returned to Germany, he received a grateful letter from Kramers. Because of Heisenberg's visit, some of the restrictions on the Dutch physicists were lifted. The Dutch Ministry of Education, which was under pressure to demonstrate cultural collaboration with German scholars, was pleased as well. Heisenberg and Kramers subsequently took advantage of the opportunity to correspond about theoretical physics far removed from practical applications or war work. When Heisenberg was asked by the Dutch Ministry of Education whether he thought that more cultural exchanges should be organised, he begged off, noting that 'when the war had come to a happy end', from the present perspective, an ambiguous statement, normal scientific exchanges would no doubt resume, but for the time being one should wait.

Did Kramers collaborate? Yes. Was this 'good' or 'bad' collaboration? Such polarised dichotomies should probably be avoided. Instead, Kramers' role in Heisenberg's visit should be placed into context and evaluated carefully. From the perspective of an outside observer, Dutch or otherwise, Kramers and his colleagues were welcoming a famous and nationalist German scientist into their institutes and labs at a time when The Netherlands were suffering under a harsh German occupation. From the perspective of an insider, a member of the Dutch physics community, it probably was understood that showing professional courtesy to a respected colleague, and thereby in no way aiding the German war effort, was an acceptable compromise with the occupying regime if it meant an improvement in their personal and professional situations. Indeed, this visit must have appeared differently to individuals with different perspectives. But from our perspective this collaboration was arguably neither 'good' nor 'bad'. It just was.

Why Comparison?

What does a comparative perspective gain? Why compare scientific research from different countries or political systems? Science is international and, in particular, international standards for publication and recognition have the effect of homogenising research, even during wartime, when lines of communication are cut and secrecy is an important consideration. When looking at scientists working in Germany during the Second World War, one naturally asks whether these researchers were motivated by support of National Socialism or its leader, Adolf Hitler. However, if Soviet, Japanese, American, British and émigré scientists did the same thing in their countries and under their corresponding political systems, then we can hardly attribute this to National Socialist convictions.

Work on nuclear weapons provides a specific example.[61] Very many scientists and scholars have asked whether the fact that the physicist Werner Heisenberg worked on nuclear weapons in Germany during the war demonstrates that he was a 'Nazi'. By this logic, one should also ask whether the Soviet physicist Igor Kurchatov was a Stalinist, or the Japanese physicist Yoshio Nishina a militarist, or the American Robert Oppenheimer a liberal democrat and capitalist. Actually, this might also be true for Oppenheimer, but the point is that historians usually do not even ask such questions except for the German case. Conversely, if it is natural that Kurchatov, Nishina and Oppenheimer have worked for their governments to develop atomic bombs, is it surprising that Heisenberg worked for the National Socialists?

Comparison helps distinguish local peculiarities from more common developments and discard some potential explanations. Local and specific explanations and contexts should not be denied or minimised, but they should also be put in the broader context. In particular, the broader the context and comparison becomes, the more the similarities appear to outweigh the differences and closer we come to determining what happens when science and war come together.

Notes

1 For a recent survey of the topic, see T. Schachtman, *Terrors and Marvels: How Science and Technology Changed the Character and Outcome of World War II* (New York: William Morrow, 2002).
2 R. Macleod (ed.), *Science and the Pacific War: Science and Survival in the Pacific, 1939–1945* (Dordrecht: Kluwer, 2000), 1–2.
3 D. Edgerton, *Warfare State: Britain, 1920–1970* (Cambridge: Cambridge University Press, 2006), 13–14.
4 See for example, P. Josephson, *Physics and Politics in Revolutionary Russia* (Berkeley: University of California Press, 1991); L. Graham, *Science in Russia and the Soviet Union* (Cambridge: Cambridge University Press, 1993); N. Krementsov, *Stalinist Science* (Princeton: Princeton University Press, 1996); A. Kojevnikov, *Stalin's Great Science: The Times and Adventures of Soviet Physicists* (London: Imperial College Press, 2004).
5 D. Holloway, *Stalin and the Bomb* (New Haven: Yale University Press, 1994).
6 D. Kevles, *The Physicists: The History of a Scientific Community in Modern America* (New York: Knopf, 1977).
7 Perhaps the best history is R. Rhodes, *The Making of the Atomic Bomb* (New York: Simon and Schuster, 1987).
8 R. Buderi, *The Invention That Changed the World: How a Small Group of RADAR Pioneers Won the Second World War and Launched a Technological Revolution* (New York: Simon and Schuster, 1996).
9 See for example, A. Thackray (ed.), 'Science after '40'', *Osiris* Vol. 7 (Chicago: University of Chicago, 1992).
10 I am limiting myself here to English language literature for Japan. See M. Low, 'Japan's Secret War? "Instant" Scientific Manpower and Japan's World War II Atomic Bomb Project', *Annals of Science*, 47 (1990), 347–60; M. Yamazaki, 'The Mobilisation of Science and Technology during the Second World War in Japan – A Historical Study of the Activities of the Technology Board Based Upon the Files of Tadashiro Inoue', *Historia Scientiarium*, 5 (1995), 167–81; M. Low, 'The Useful War: Radar and the Mobilisation of Science and Industry in Japan', Macleod 2001, 291–301; K. Nagase-Reimer, W. Grunden, and M. Yamazaki, 'Nuclear Weapons Research in Japan during the Second World War', *Historia Scientiarium*, 14, (2005), 201–40; M. Low, *Science and the Building of a New Japan* (London: Palgrave Macmillian, 2005); W. Grunden, Y. Kawamura, E. Kolchinsky and H. Maier, 'Laying the foundation for wartime research: A comparative overview of science mobilisation in National Socialist Germany, Japan, and the Soviet Union', C. Sachse and M. Walker (eds), *Politics and Science in Wartime: Comparative International Perspectives on the Kaiser Wilhelm Institutes, Osiris* Vol. 20 (Chicago: University of Chicago Press, 2005), 79–106; W. Grunden, M. Walker and M. Yamazaki, 'Wartime Nuclear Weapons Research in Germany and Japan', Sachse and Walker 2005, 107–30.

11 W. Grunden, *Secret Weapons and World War II: Japan in the Shadow of Big Science* (Lawrence: University of Kansas, 2005).

12 K.-H. Ludwig, *Technik und Ingenieure im Dritten Reich* (Düsseldorf: Droste Verlag, 1974).

13 H. Trischler, *Luft- und Raumfahrtforschung in Deutschland 1900–1970. Politische Geschichte einer Wissenschaft* (Frankfurt/Main: Campus, 1992); M. Neufeld, *The Rocket and the Reich: Peenemünde and the Coming of the Ballistic Missile* (New York: The Free Press, 1995); M. Walker, *German National Socialism and the Quest for Nuclear Power, 1939–1949* (Cambridge, Cambridge University Press, 1989).

14 See for example, H. Mehrtens, 'Angewandte Mathematik und Anwendungen der Mathematik im nationalsozialistischen Deutschland', *Geschichte und Gesellschaft*, 12 (1986), 317–47; H. Maier, *Erwin Marx (1893–1980), Ingenieurwissenschaftler in Braunschweig, und die Forschung und Entwicklung auf dem Gebiet der elektrischen Energieübertragung auf weite Entfernungen zwischen 1918 und 1950* (Stuttgart: GNT, 1993); M. Renneberg and M. Walker (eds), *Science, Technology, and National Socialism* (Cambridge: Cambridge University Press, 1993); U. Deichmann, *Biologists under Hitler* (Cambridge, MA: Harvard University Press, 1996); U. Deichmann, *Flüchten, Mitmachen, Vergessen. Chemiker und Biochemiker in der NS-Zeit* (Weinheim: VCH, 2001). Publications that do emphasise the war include the following: H. Mehrtens, 'Mathematics and War: Germany 1900–1945', J.M. Sanchez-Ron and P. Forman (eds), *National Military Establishments and the Advancement of Science and Technology: Studies in Twentieth Century History* (Dordrecht: Kluwer Academic Publishers, 1996), 87–134; U. Deichmann, 'Kriegsbezogene biologische, biochemische und chemische Forschung an den Kaiser Wilhelm-Instituten für Züchtungsforschung, für Physikalische Chemie und Elektrochemie und für Medizinische Forschung', D. Kaufmann (ed.), *Geschichte der Kaiser-Wilhelm-Gesellschaft im Nationalsozialismus. Bestandsaufnahme und Perspektiven der Forschung. Volume 1* (Göttingen: Wallstein, 2000), 231–57; M. Epple and V. Remmert, '"Eine ungeahnte Synthese zwischen reiner und angewandter Mathematik": Kriegsrelevante mathematische Forschung in Deutschland während des II. Weltkrieges', Kaufmann 2000, Vol. 1, 258–95; M. Eckert, 'Theoretische Physiker in Kriegsprojekten. Zur Problematik einer internationalen vergleichenden Analyse', Kaufmann 2000, Vol. 1, 296–308; H. Maier (ed.), *Rüstungsforschung im Nationalsozialismus. Organisation, Mobilisierung und Entgrenzung der Technikwissenschaften* (Göttingen: Wallstein, 2002); F. Schmaltz, *Kampfstoff-Forschung im Nationalsozialismus. Zur Kooperation von Kaiser Wilhelm Instituten, Militär und Industrie* (Göttingen: Wallstein, 2005).

15 Maier 2002; S. Heim (ed.), *Autarkie und Ostexpansion. Pflanzenzucht und Agrarforschung im Nationalsozialismus* (Göttingen: Wallstein, 2002); C. Sachse (ed.), *Die Verbindung nach Auschwitz. Biowissenschaften und Menschenversuche an Kaiser-Wilhelm-Instituten. Dokumentation eines Symposiums* (Göttingen: Wallstein, 2003); H.-W. Schmuhl (ed.), *Rassenforschung an Kaiser-Wilhelm-Instituten vor und nach 1933* (Göttingen: Wallstein, 2003); W. Schieder and A. Trunk (eds), *Adolf Butenandt und die Kaiser-Wilhelm-Gesellschaft. Wissenschaft, Industrie und Politik im 'Dritten Reich'* (Göttingen: Wallstein, 2004); H. Maier, *Forschung als Waffe. Rüstungsforschung in der Kaiser-Wilhelm-Gesellschaft und das Kaiser-Wilhelm-Institut für Metallforschung, 1900–1945/48* (Göttingen: Wallstein, 2007).

16 Unfortunately, whatever Italian-language literature that exists is not accessible to this author; however, see M. Epple, A. Karachalios and V. Remmert, 'Aerodynamics and Mathematics in National Socialist Germany and Fascist Italy: A Comparison of Research Institutes', Sachse and Walker 2005, 131–58.

17 See N. Chevassus-Au-Louis, *Savants sous l'occupation: Enquête sur la vie scientifique entre 1940 et 1944* (Paris: Le Seuil, 2006).

18 For the émigrés, see M. Ash and A. Söllner, 'Forced Migration and Scientific Change after 1933', M. Ash and A. Söllner (eds), *Forced Migration and Scientific Change:*

Emigré German-Speaking Scientists and Scholars after 1933 (Cambridge, Cambridge University Press, 1996), 1–19; R. Siegmund-Schultze, *Mathematiker auf der Flucht vor Hitler. Quellen und Studien zur Emigration einer Wissenschaft* (Wiesbaden: Vieweg, 1998); M. Ash, 'Emigration und Wissenschaftswandel als Folgen der nationalsozialistischen Wissenschaftspolitik', D. Kaufmann (ed.), *Geschichte der Kaiser-Wilhelm-Gesellschaft im Nationalsozialismus. Bestandsaufnahme und Perspektiven der Forschung, Volume 2* (Göttingen: Wallstein, 2000), 610–31; S. Wolff, 'Die Ausgrenzung und Vertreibung der Physiker im Nationalsozialismus-welche Rolle spielte die Deutsche Physikalische Gesellschaft?' D. Hoffmann and M. Walker (eds), *Physiker zwischen Autonomie und Anpassung – Die DPG im Dritte Reich* (Weinheim: VCH, 2007), 91–138.

19 See the chapters in M. Walker (ed.), *Science and Ideology: A Comparative History* (London: Routledge, 2003) and Sachse and Walker 2005.

20 I would like to thank Ad Maas for bringing this point to my attention.

21 See M. Walker, *Nazi Science: Myth, Truth, and the German Atom Bomb* (New York, Perseus Publishing, 1995), 130–8.

22 Hoffmann and Walker 2007, 185–215.

23 Grunden, Kawamura, Kolchinsky, Maier and Yamazaki 2005.

24 P. Wegener, *The Peenemunde Wind Tunnels: A Memoir* (New Haven: Yale University Press, 1996).

25 This passage draws upon a collaborative article, M. Gordin, W. Grunden, M. Walker and Z. Wang '"Ideologically Correct" Science', Walker 2003, 35–65; this article also examines science during the French Revolution, the McCarthy period in the United States, and the Chinese communist Cultural Revolution.

26 For example, see D. Hoffmann and M. Walker, 'Friedrich Möglich: A Scientist's Journey from Fascism to Communism', Walker 2003, 227–60.

27 See Krementsov 1996, Kojevnikov 2004.

28 On Oppenheimer, see K. Bird and M.J. Sherwin, *American Prometheus: The Triumph and Tragedy of J. Robert Oppenheimer* (New York: Knopf, 2005); D. Cassidy, *J. Robert Oppenheimer: And the American Century* (New York: PI Press, 2005).

29 Walker 1989, 229–33.

30 S. Harris, *Factories of death: Japanese biological warfare 1932–45 and the American cover-up* (Routledge, London, 1994); H. Gold, *Unit 731 testimony* (Boston: Charles E Tuttle, 1996); K. Tsuneishi, *Secrets of Unit 731* (Boulder: Rowman and Littlefield, forthcoming).

31 A. Mitscherlich and F. Mielke, *The Death Doctors* (London: Elek, 1962) as well as R. J. Lifton, *The Nazi Doctors: Medical Killing and the Psychology of Genocide* (New York: Basic, 1986).

32 See Sachse 2003.

33 See G. Baader, S. Lederer, M. Low, F. Schmaltz and A. von Schwerin, 'Pathways to Human Experimentation, 1933–1945: Germany, Japan, and the United States', in Sachse and Walker 2005, 205–31.

34 See A. von Schwerin, *Experimentalisierung des Menschen. Der Genetiker Hans Nachtsheim und die vergleichende Erbpathologie, 1920–1945* (Göttingen: Wallstein, 2004).

35 Neufeld 1995.

36 U. Albrecht, 'Military Technology and National Socialist Ideology', Renneberg and Walker 1993, 88–125; 358–63.

37 Walker 1989.

38 For particle beam research, see B. Weiss, *'Forschungsstelle D'. Der Schweizer Ingenieur Walter Dällenbach (1892–1990), die AEG und die Entwicklung kernphysikalischer Grossgeräte im nationalsozialistischen Deutschland* (Berlin: Michael Engel, 1996).

39 For chemical weapons, see Schmaltz 2005.
40 R. Karlsch, *Hitlers Bombe* (Munich: DVA, 2005); R. Karlsch and M. Walker, 'New Light on Hitler's Bomb', *Physics World*, (June, 2005), 15–18.
41 G. Altschuler, 'The Convictions of Peter Debye', *Daedalus*, 135/4 (2006), 96–103; D. Hoffmann, 'Peter Debye (1884–1966). Ein Dossier', *Max-Planck-Institut für Wissenschaftsgeschichte Preprints, 314* (Berlin: MPIfWG, 2006).
42 Several contributors to this volume should be commended for directly addressing this sensitive topic.
43 Unfortunately, much less is known about industrial science because of the difficulty of getting access to archival sources, but it is a question for further inquiry.
44 Samuel Goudsmit to Vannevar Bush (23 August 1945) Samuel Goudsmit Papers, American Institute of Physics (SGP).
45 James Franck to Samuel Goudsmit (6 December 1941) SGP.
46 S. Goudsmit, *Alsos*, 2nd edn. (Los Angeles: Tomash, 1983), 84.
47 Goudsmit 1983, 161.
48 Goudsmit 1983, 235.
49 Goudsmit 1983, 80.
50 Goudsmit 1983, 102.
51 Goudsmit 1983, 121.
52 Goudsmit 1983, 112–13.
53 Goudsmit 1983, 35.
54 Goudsmit 1983, 104.
55 J. Haberer, *Politics and the Community of Science* (New York: van Nostrand, 1969), 141–2.
56 Haberer 1969, 150.
57 Haberer 1969, 153.
58 A. Beyerchen, *Scientists under Hitler: Politics and the Physics Community in the Third Reich* (New Haven: Yale University Press, 1977); H. Mehrtens and S. Richter (eds), *Naturwissenschaft, Technik, und NS-Ideologie: Beiträge zur Wissenschaftsgeschichte des Dritten Reichs* (Frankfurt am Main: Suhrkamp, 1980).
59 Walker 1995, 166–70.
60 Walker 1995, 169.
61 M. Walker, 'A Comparative History of Nuclear Weapons', Kaufmann 2000, vol. 1, 309–27.

2 To work or not to work in war research?

The case of the Italian physicist G. P. S. Occhialini during World War II

Leonardo Gariboldi

The small community of physicists in Italy was not active in war research during the Second World War. A number of factors – fewer physicists because of emigration due to Italy's racial laws, their unwillingness to give active support to the Fascist government and its war policies, the general economic situation – reduced the amount of research in physics in Italy. While physicists in Italy continued with their scientific research, many Italian physicists abroad instead had to face the choice of whether or not to work for war research in the Allied countries. The case of the Italian physicist Giuseppe (better known by his nickname 'Beppo') Paolo Stanislao Occhialini (1907–1993)[1] is an interesting one because of the changing conditions that affected his choices.

The Italian Context

In 1930, ten years before Italy entered the Second World War, two Italian physics research groups, located in Rome and Florence, were playing an important role at an international level.[2] Scientists working in these groups were actually involved in some of the most important research in the physical sciences of the time.

The Roman group,[3] the *ragazzi di via Panisperna*, was mainly concerned with the study of nuclear and neutron physics,[4] while the Florentine group,[5] the *Arcetri school*, focused on cosmic radiation.[6] Both groups developed technical instrumentation, such as the Rossi coincidence circuit, that had a profound impact on the future development of physics the world over. The know-how of the researchers of both the Roman and the Florentine groups might have been used over the next decade in war research, and this did indeed happen to some of the people concerned, but not in Italy.

The Rome group came to an abrupt halt when Enrico Fermi left Italy on 6 December 1938 to go to Stockholm for the Nobel Prize ceremony. From Stockholm Fermi went to Columbia University with the permission of the Fascist government to give a physics course. Because of the racial laws[7] which affected Fermi's Jewish wife, Fermi decided to leave Italy and not to come back after the end of his course at Columbia University.[8]

Fermi's colleague Emilio Segrè was already in Berkeley at the beginning of the summer of 1938, working on the radioisotopes of technetium, a chemical element he had discovered in Palermo. As a Jew, he decided not to come back to Italy and called his family to the United States.[9] A third Roman physicist, Bruno Pontecorvo, had already moved to the *Institut du Radium* in Paris in order to study and work with Frédéric Joliot.[10] Of the other members of the Roman group, Franco Rasetti, disagreeing with the racial laws, also chose to go abroad (to the Laval University in Canada) without being compelled by these laws or other political factors.

The Florence group dissolved completely even before the racial laws. Most of its members continued to work in Italy during the mid-1930s, such as Gilberto Bernardini, who moved to Bologna. Its scientific leader, Bruno Rossi, initially moved to the Institute of Physics at the University of Padua, but, suffering from the racial laws, eventually had to leave the country altogether. He left Italy for Copenhagen on 12 October 1938. He then moved to the United Kingdom and to the United States, where he worked on the Manhattan Project. Giulio Racah left Italy for Palestine, where he contributed to the studies of theoretical physics at the recently founded Hebrew University of Jerusalem. Beppo Occhialini, as we will see, ended up in Cambridge.

When Italy entered the Second World War in 1940, the Roman group, formally directed by Antonino Lo Surdo, acquired a new scientific leader, Edoardo Amaldi. Amaldi tried and mostly succeeded in gathering most of what was left

Figure 2.1 The research group in São Paulo. From left to right: Roberto Xavier de Oliveira, Dona maria Caseira, Giuseppe Occhialini, Marcelo Damy de Souza Santos, José Caseiro, Yolande Monteux, Abrahão de Moraes, Mário Schenberg, Gleb Wataghin, Guidolino Bentivoglio.

of Italian physics in Rome.[11] The main fields of research were again the two 'traditional' ones studied in Rome and Florence, i.e. nuclear (neutron) physics and cosmic ray physics.

The other Italian universities had much smaller groups of physicists, not all of them devoted to physical research but often just to teaching physics. In Milan[12] research in cosmic ray physics was done by Giuseppe Cocconi with the support of Giovanni Polvani, while Giovanni Gentile Jr formulated the intermediate statistics or parastatistics (i.e. intermediate between the Fermi-Dirac and the Bose-Einstein statistics). There were other small groups of physicists in Bologna (Gilberto Bernardini, Quirino Majorana), Genoa (Augusto Occhialini), Naples (Antonio Carrelli), Padua (Nicolò Dallaporta, Antonio Rostagni), Palermo (Mariano Santangelo), Pavia (Piero Caldirola, Luigi Giulotto) and Turin[13] (Enrico Persico).

After the armistice, signed by the Italian government in September 1943, German troops invaded northern and central Italy. A puppet state with a Fascist government was created in the German-controlled regions of Italy, the Italian Social Republic or Republic of Salò.

To avoid damage due to Allied bombing and to avoid the requisition of scientific instruments by the German troops, most experiments were interrupted and the dismantled instrumentation was hidden in the countryside. One of the few exceptions was Conversi, Pancini and Piccioni's experiment in Rome – one of the most important European experiments in physics during and soon after the war – which continued in a hidden laboratory in a high school next to the Vatican State. This location granted a zone of protection from the bombing.

Italian physicists in Italy generally refused to work in war research. In particular, to avoid any compulsory collaboration with Nazi war research, the Roman physicists stopped all research in nuclear physics in 1941.

According to Occhialini, the Italian physicists in Rome had to face a 'problem of choice':

> you have to choose for the present or for the future. If it is the present, the thing is simple: you join an army of any kind, official or clandestine, but you join it. If not, you decide that the best thing is to write a book, meaning that this book is going to substitute your action and you devote yourself to prepare the future of young people.[14]

The Roman physicists decided that the best thing to do was to deploy research in cosmic ray physics to educate the next generation of physicists. In Occhialini's opinion, the racial laws spurred the physicists in Rome to take up this task. When Occhialini came back to Italy in late 1938 for a short holiday he felt that something had changed. He perceived the existence of a popular, hidden 'revolution' of the Italian people against the expulsion of the Jews that was in effect the first step in the re-conquest of their own Italian soul.[15]

On the other hand, most Italian physicists abroad agreed to be involved in war research. In the United States Enrico Fermi, Emilio Segrè, Bruno Rossi and Sergio De Benedetti worked on the Manhattan project. Bruno Pontecorvo, a Jew, had to leave France and escaped to the United States, where he studied neutronic probing in the oil fields of Oklahoma. In 1943–1944 he worked in a research group on nuclear reactors in Canada with Pierre Auger, but he could not keep in touch with his colleagues on the Manhattan project on secret issues. Franco Rasetti refused to join war research.

The case of Beppo Occhialini is maybe the least simple. During the war he had to face different personal situations, which implied different 'choices'. When offered the possibility of joining war research in Brazil, he chose not to, while when he offered to join war research in England, he was not chosen.

The First Choice: Not to Work for War Research

Occhialini had joined the Fascist Party in 1922 when he was 15 years old (probably in adolescent opposition to his father's liberal ideas). He changed his mind when a member of Parliament, Giacomo Matteotti, was killed by the Fascists in 1924. In the mid-1930s he was active in printing clandestine propaganda against the regime, in particular on the occasion of the 1935 Ethiopian and 1936 Spanish wars. To avoid possible imprisonment, in 1937 Occhialini accepted an invitation from the Ukrainian-born Italian physicist Gleb Vassilievich Wataghin (1899–1986) to join his research group in São Paulo.

So, when the Second World War began in 1939, Occhialini was in São Paulo as a scientific leader of a research group on cosmic rays at the Institute of Physics.[16] The group comprised, among others, Marcelo Damy de Souza Santos[17] and Mário Schönberg (or Schenberg).[18] Occhialini's research concerned cosmic rays and radioactivity as well as the development of devices used to study radiation (cloud chamber, Geiger-Müller counter, and other kinds of counters).[19]

In 1940, Occhialini was looking for a way to go to the United States to work in cosmic ray physics. His preferences were Chicago, where he aimed to work with Arthur Holly Compton and Bruno Rossi, and Pasadena, to work with Robert Andrews Millikan. Another interesting field of research in the United States was the study of neutron radiation in a laboratory with a cyclotron or high-tension tubes, such as in Berkeley with Ernest Orlando Lawrence or in Washington with Merle Antony Tuve. If a fellowship had been granted, his work might have favoured his possible engagement on the Manhattan project.[20] However, he was not granted a fellowship.

Even though Occhialini was not able to go to the United States, he worked closely with Compton on a campaign organised in Brazil by Wataghin of measurements on cosmic rays, mainly the latitude effect. Their results were communicated at an international symposium held in Rio de Janeiro on 4–8 August 1941.[21]

When Brazil joined the Allied forces in March 1942, the Physics Laboratory of the University of São Paulo began to work on war research. The best experimental physicist in Brazil, Marcelo Damy de Souza Santos, became director of the Institute of Physics and of the laboratory itself. De Souza Santos decided to abandon his research in cosmic ray physics and to engage in war research, mainly to study detectors for German U-boats. The part of the laboratory led by Damy de Souza

Santos worked on war research, 'working for the present', while Wataghin, who, as an Italian citizen, was obliged to resign as director of the Institute, went on working with young people for their 'future education'.

After Brazil had entered the war, the life of Italian citizens in Brazil worsened. They had to travel with a special type of internal passport and they were forbidden to speak Italian or any other foreign language. Since Occhialini was now formally an enemy alien, he had to face the problem of working in a laboratory where some people were doing war research against Italy. According to Damy de Souza Santos, it was a common habit in the laboratory not to talk about politics.

> Wataghin did not allow any political discussion. People lived in a purely scientific world. The only concern was for science.[22]

The outbreak of war compelled the physicists in São Paulo to face a political problem. Occhialini, in order to avoid repercussions of his anti-fascism among his colleagues, decided that the best thing to do was to leave the laboratory and even São Paulo altogether. He was confident in Damy de Souza Santos and had no wish, whatever might happen, for Damy de Souza Santos to get into trouble because of him:

> Leaving São Paulo completely was the only thing I could do *to leave the people free to work without a suspect*. [...] I envisaged the possibility that, at any time, a suspect, sabotage in the laboratory, could make them responsible because of me.[23]

Although Occhialini might have continued to study cosmic rays with Wataghin, doing purely scientific non-war research, he preferred to stop. Occhialini decided to exile himself in the Itatiaia Mountains, between São Paulo and Rio de Janeiro, where he lived as an alpine guide.

The Second Choice: To Work for War Research

When Italy signed the armistice in September 1943 Occhialini came out of exile. He went to Rezende and then to Rio de Janeiro to check on the possibility of going to Italy with the utmost urgency.

Back in Rio de Janeiro, Occhialini was under the protection of Carlos Chagas Jr (1910–2000) at the Biophysics Laboratory.[24] Chagas met marshal Eurico Gaspar Dutra, then the Brazilian Minister of War, and suggested that Occhialini join the *Força Expedicionaria Brasileira* [FEB], the Brazilian troops sent to fight in the Mediterranean, as an ideal official of delegation because of his good knowledge of the English, French, Italian and Portuguese languages. Chagas' suggestion was rejected by Dutra. Occhialini later remarked that his presence in the FEB would have been quite anomalous, because most of the 25,000 or so people joining it were very young and not prepared to fight a war.[25]

Eventually, Occhialini would go not to Italy but to England. From February 1943 his English colleague and friend, Patrick Blackett,[26] then at the British Admiralty, had been trying to help him in case he had been interned.[27] Blackett asked the Foreign Office to discover Occhialini's whereabouts through the British Embassy in Brazil, but because of Occhialini's isolation in the mountains of Itatiaia it was very difficult, if not impossible, to find information about him. According to the British Embassy in Rio, the Itamarati (the Brazilian Ministry of Exterior) had stopped Occhialini from returning to Italy in July 1942

> […] *because of the importance of his work for scientific warfare.* Officially he was made to resign and went on holiday but we understand that he will be resuming his former activities in an unofficial capacity in the course of the next few weeks.[28]

In October 1943 the Foreign Office received information that Occhialini was anxious to collaborate with the Allied cause,[29] troubled as he was by the idea that the Germans could make a nuclear weapon before the Allied scientists.

> Notwithstanding the absence of any contact, I knew that the atomic bomb was possible. I thought that the Germans were building it, and I hoped that, on the other side, [the Allies] were doing the same.[30]

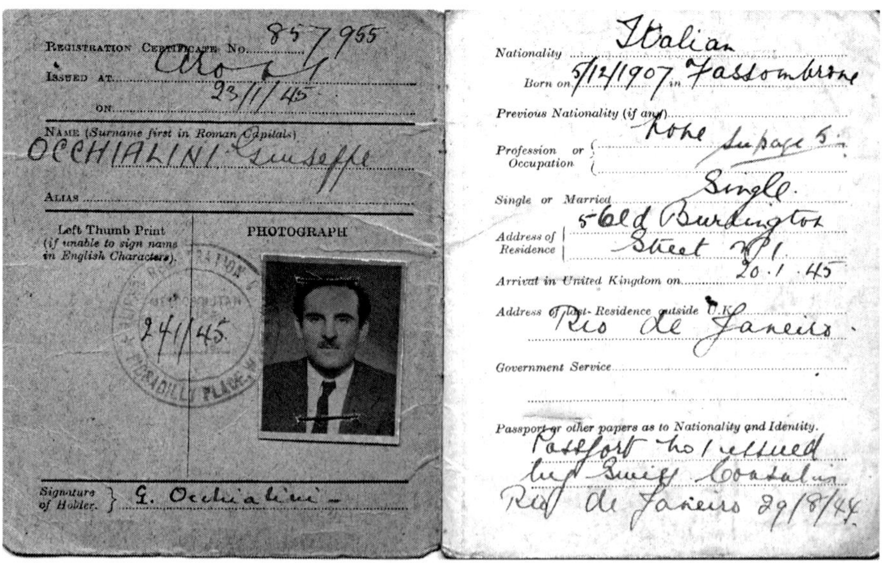
Registration Certificate No. 857955
Issued at
On 23/1/45
Name (Surname first in Roman Capitals) OCCHIALINI Giuseppe
Alias
Left Thumb Print (if unable to sign name in English Characters).
PHOTOGRAPH
24/1/45
Signature of Holder. G. Occhialini
Nationality Italian
Born on 5/12/1907 in Fossombrone
Previous Nationality (if any) None
Profession or Occupation
Single or Married Single.
Address of Residence 5 Old Burlington Street W1
Arrival in United Kingdom on 20.1.45
Address of last Residence outside U.K. Rio de Janeiro.
Government Service
Passport or other papers as to Nationality and Identity.
Passport No / issued by Swiss Consulate Rio de Janeiro 29/8/44.

Figure 2.2 Registration Certificate issued to Occhialini after his arrival in the United Kingdom

The documentation at our disposal prevents us from knowing Occhialini's actual intentions in more detail. His willingness to join an army fighting for Italy's freedom and his anxiety to collaborate as a scientist, in England or anywhere else, on the Allies' war projects were probably two options that he had to face constantly, while waiting for the first actual opportunity, although the former is the prevailing one in his *a posteriori* comments.

The Foreign Office started correspondence with Blackett in order to gather information about Occhialini's qualities as a physicist and his political sympathies.[31] According to Blackett:

> Occhialini has always been both definitely and at times actively anti-Fascist. As far as I know at no time has he ever cooperated willingly with the Fascist authorities. [...] He has one other qualification of note – he has a marked adventurous character and is a distinguished speliologist [*sic*] and I believe holds nearly the world's record for depth of descent into caves. *I feel very strongly this combination of physics and guts might be usefully utilised.* I would like to add that I would be personally delighted to see him made use of in any possible way.[32]

A similar answer was given by Blackett to Lord Rothschild[33] when the War Office started to analyse Occhialini's possible involvement in war activities. Lord Rothschild had written to Blackett asking for information on Occhialini:

> *Presumably his work* might *be concerned with atomic physics, much of which is particularly secret at the moment.* Before agreeing to his coming, the Security Authorities are naturally anxious to know something about his political sympathies. I should be very grateful if you could let me know what you think about this. I imagine he is anti-Fascist, but you will doubtless be able to say, having worked with him.[34]

Fearful of in any way harming his father, an important professor of physics and part of a family with a tradition of liberal political thought, Occhialini asked the British institutions to keep correspondence and plans on him secret.

> As Occhialini's parents are in occupied Italy, he requests that his application should be treated as strictly secret and confidential.[35]

Permission to go to England was granted in February 1944. Since Occhialini was supposed to leave Brazil at any moment, he refused to accept a chair of physics at the University of São Paulo again.[36] For the same reason, he also declined an offer from Theodor J. Wang of Ohio State University to join him in his research on the 4 MeV bevatron that was being built by him and Lorenzo Emo Capodilista (a former colleague of Occhialini in Florence).[37]

Before leaving Brazil for Europe Occhialini taught again at the University of São Paulo.[38] In October 1944, he gave lectures in a course on X-rays for postgraduate students.[39] It was on this occasion that he met Cesare Lattes and Ugo Camerini, two promising young students who wished to work with a cloud chamber. Lattes and Camerini later joined Occhialini in Bristol to work with Powell's group on nuclear emulsions.

Occhialini left Brazil for Europe in November 1944. Notwithstanding the fact that his eventual occupation was still undecided, he landed in England in January 1945 hoping to fight.[40]

> When I got in touch with the English, I wished to be a parachutist and I started to train. I never contemplated the possibility of returning to physics or any other thing. It was practically a kind of suicide. But when I left for England I discovered that the English were not interested in me as a parachutist or for war research [...].[41]

Employment on a special project under the Department of Scientific and Industrial Research, as was intended initially, proved instead to be impossible. Occhialini then attempted to join the war effort with references from Sir Edward Appelton and Patrick Blackett.

> The Aliens War Service Department has been informed of Signor Occhialini's presence in this country. *He is anxious to undertake any work to assist the war effort, though it is of course to be preferred that his special talents should be made use of.* [...] If it were proposed to employ him on secret work, formal security clearance would, of course, have to be obtained, but we do not think that there is likely to be any difficulty about this.[42]

In the meanwhile, Occhialini spent a short time in Manchester and in Cambridge. In Manchester he visited the laboratory directed by Blackett. Occhialini did not ask Blackett to join the Manchester group because he did not want to take advantage of the fact that few physicists were working in the physics laboratories in England.

In Cambridge, William Lawrence Bragg, who was willing to help Occhialini, suggested a possible alternative to laboratory work: collaboration with the British Council. Bragg gave Occhialini a note to F. J. R. Bottrall, who was going to be sent to Italy for the British Council. Occhialini might help Bottrall by giving him information about the Italian scientists.

> A job of this kind, it seems to me, would be just what he is looking for. [...] Failing this, I might try to find him a job here to help one of our teams, but I feel this is rather a second-best solution. *The sort of work going on in the Laboratory is ceasing to help the war directly now that the end is approaching.*[43]

The most serious attempt to employ Occhialini during the last months of the war was that of the General Electric Company. In March 1945, it put an application through the Aliens War Service Department for permission to employ him on secret work.[44] In particular, Wallace Akers, the director of the Department of

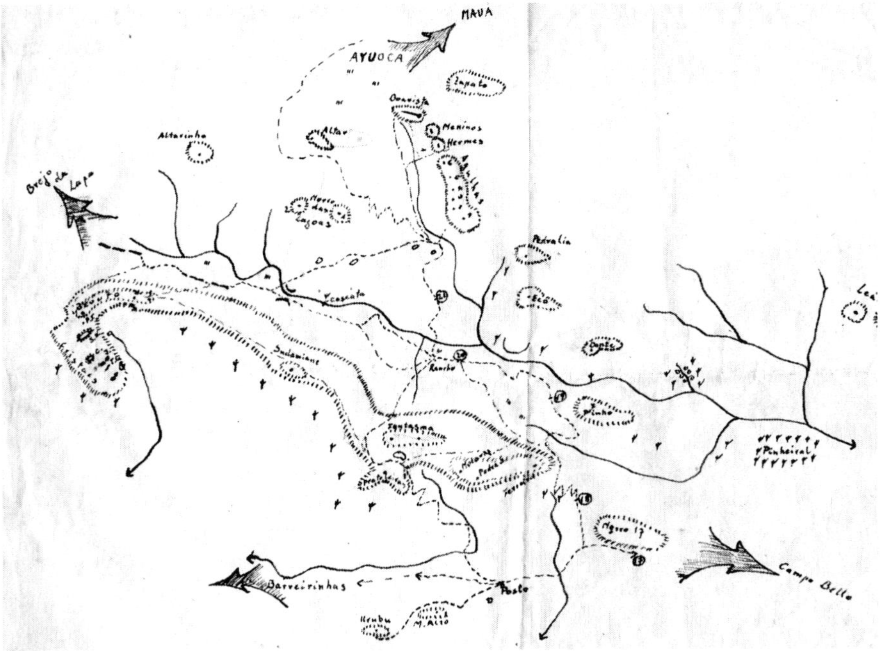

Figure 2.3 A map of the Itatiaia Mountains drawn by Occhialini during his exile in 1942–1943.

Scientific and Industrial Research, suggested that Occhialini work on the Tube Alloys, the code name for the British atomic bomb effort, and on some of the crystal work under the Coordination of Valve Development. Unfortunately for him, this last attempt failed. Notwithstanding the fact that, with Blackett's reference, the security question seemed cleared up,[45] the Aliens War Service Department informed the General Electric Company that it was unable to grant Occhialini a permit for his employment at its research laboratories.[46]

The war in Europe was almost at an end. No further attempt to join war research was supported. Blackett himself now helped Occhialini to return to pure scientific research, this time in Bristol, at the Wills Laboratories, where Occhialini joined Cecil Frank Powell (1903–1969) and his group working on nuclear emulsions. The most important scientific result of their collaboration would be the discovery of the pion.[47]

Conclusion

In the difficult years of the Second World War Occhialini twice had to face a choice that he thought also concerned his colleagues who remained in Italy: to work for the present or to work for the future? Working for war research (or joining an army) or educating young people?

> The point of view which I tried to offer is about choice: between making research or making war, between being a terrorist or being a scientist, between fighting against dictatorship or working in a laboratory.[48]

Occhialini made his choices for political reasons, which consequently affected his scientific career. The first choice – leaving São Paulo and going into exile in the Itatiaia mountains – was made to free his Brazilian colleagues to do war research without suspicion or even sabotage because of his presence. It was a paradoxical decision not to do scientific research in the present in order to make scientific research continue for the future. The second choice was based on the more general desire to help his own country to be free again, even working in war research if necessary. His desire, however, ran up against the decisions of various British institutions. As the war neared its end, the probability of his occupation in war research decreased and, eventually, he was not engaged in any such research.

> I left Brazil, and the whole world told me it was silly of me; I dreamt of coming back to Italy on a tank, as a conqueror; I dreamt in vain of doing war research. *None of this happened.*[49]

Aknowledgements

I would like to thank Amélia Império Hamburger and her staff (Universidade de São Paulo [USP]), the Wataghin Archive (USP), the Royal Society of London and the Occhialini-Dilworth Archive (Università degli Studi di Milano). Heartfelt thanks to Marcelo Damy de Souza Santos for his conversation with me about his collaboration with Occhialini in São Paulo.

Abbreviations

BP Royal Society. Blackett Papers.

CBD Sociedade Brasileira para o Progresso da Ciência, Cientistas do Brasil. Depoimentos, São Paulo: SBPC, 1998.

ODA University of Milan. Occhialini-Dilworth Archive.

SLBO The Scientific Legacy of Beppo Occhialini, edited by P. Redondi, G. Sironi, P. Tucci and G. Vegni, Bologna: SIF – Springer Verlag, 2006.

WA University of São Paulo. Wataghin Archive.

WAIO Interview with Occhialini (São Paulo, 8 October 1982) by Cecil Robilotta, Manoel Robilotta, João Zanetic, M. Regina Kawamura, Alexandre Mederios, Ernst Hamburger, Amélia Império Hamburger. Unpublished. The copy quoted is kept unclassified in the Wataghin Archive, São Paulo.

Notes

1 On Beppo Occhialini, see: L. Gariboldi and P. Tucci, 'Giuseppe Paolo Stanislao Occhialini (1907–1993). A Short Biography', SLBO, pp. xi–xxxviii. More biographical and bibliographical information on Occhialini can be found in the other historical chapters in SLBO.

2 On the history of physics in Italy between the two World Wars, see: E. Amaldi, 'Italy between the Two World Wars', *Colloque International sur l'Histoire de la Physique des Particules, Journal de Physique*, Colloque C-8, supplement au n° 12, Tome 43 (1982), C8–C168; S. Galdabini and G. Giuliani, 'Physics in Italy between 1900 and 1940: The Universities, Physicists, Funds, and Research', *Historical Studies in the Physical and Biological Sciences* 19 (1988), 115–36. See also: A. Russo, 'Science and Industry in Italy between the Two World Wars', *Historical Studies in the Physical and Biological Sciences* 16 (1986), 281–320. On the relation of physicists with the Fascist government, see: G. Battimelli and M. De Maria, 'La fisica', R. Simili and G. Paoloni (eds), *Per una storia del Consiglio Nazionale delle Ricerche* vol. 1 (Roma: Laterza, 2001), 281–311; H. Kragh, *Quantum Generations. A History of Physics in the Twentieth Century* (Princeton: Princeton University Press, 1999), 238–40.

3 They were: Edoardo Amaldi (1908–1998), Oscar D'Agostino (1901–1975), Enrico Fermi (1901–1954), Ettore Majorana (1906–disappeared in 1938), Bruno Pontecorvo (1913–1993), Franco Rasetti (1901–2001), Emilio Segrè (1905–1989).

4 On the Roman group, see: E. Amaldi, 'From the discovery of the neutron to the discovery of nuclear fission', *Physics Report* 111 (1984), 1–332; E. Amaldi, 'Neutron work in Rome in 1934–36 and the discovery of uranium fission', *Rivista di Storia della Scienza* 1 (1984), 1–24; E. Amaldi, *Da via Panisperna all'America* (Roma: Editori Riuniti, 1997); F. Cordella, A. De Gregorio and F. Sebastiani, *Enrico Fermi: gli anni italiani* (Roma: Editori Riuniti, 2001); E. Segrè, 'Nuclear Physics in Rome', R. H. Stuewer (ed.) *Nuclear Physics in Retrospect. Proceedings of a Symposium on the 1930s* (Minneapolis: University of Minnesota Press, 1979), 35–62.

5 They were: Gilberto Bernardini (1906–1995), Daria Bocciarelli, Lorenzo Emo Capodilista (1909–1973), Giuseppe Occhialini (1907–1993), Giulio Racah (1909–1965), Bruno Rossi (1905–1993), with Enrico Persico (1900–1969) as professor of theoretical physics.

6 On the contribution to physics of the Arcetri school and Bruno Rossi in particular, see: A. Bonetti and M. Mazzoni, 'The Arcetri School of Physics', SLBO, 3–34; M. C. Bustamante, 'Bruno Rossi au début des années trente: une étape décisive dans la physique des rayons cosmiques', *Archives internationals d'histoire des sciences* 44 (1994), 92–115; M. De Maria, G. Malizia and A. Russo, 'La nascita della fisica dei raggi cosmici in Italia e la scoperta dell'effetto Est-Ovest', *Giornale di Fisica* 33 (1992), 207–28; M. Mandò, 'Notizie sugli studi di Fisica (1859–1949)', *Storia dell'Ateneo Fiorentino. Contributi di Studio* (Firenze: Parretti Grafiche, 1986). A short biography of Bruno Rossi is: G. W. Clark, 'Bruno Benedetto Rossi', *Biographical Memoirs of the National Academy of Sciences* 75 (1998), 310–41. Rossi's autobiography is: B. Rossi, *Momenti nella vita di uno scienziato* (Bologna: Zanichelli, 1987).

7 On racism in Italy, see: R. Maiocchi, *Scienza italiana e razzismo fascista* (Firenze: La Nuova Italia, 1999). On Italy's racial laws and physics, see: E. Amaldi, 'Il caso della fisica', *Conseguenze culturali delle leggi razziali in Italia* (Roma: Accademia Nazionale dei Lincei, 1990), 107–33.

8 On Enrico Fermi's emigration and his scientific activity in the United States, see: L. Fermi, *Atoms in the Family. My Life with Enrico Fermi* (Chicago: University of Chicago Press, 1954); G. Maltese, *Enrico Fermi in America. Una biografia scientifica: 1938–1954* (Bologna: Zanichelli, 2003); E. Segrè, *Enrico Fermi: Physicist* (Chicago: University of Chicago Press, 1970).

9 On Segrè and his scientific activity during WWII, see: E. Segrè, *A Mind Always in Motion. The Autobiography of Emilio Segrè* (Berkeley: University of California Press, 1993).

10 On Pontecorvo and his scientific activity during WWII, see the biographical chapters in: S. M. Bilenky, T. D. Blochintseva, I. G. Pokrovskaya and M. G. Sapozhnikov (eds), *B. Pontecorvo Selected Works. Recollections on B. Pontecorvo* (Bologna: SIF, 1997).

11 Besides Amaldi and Lo Surdo themselves, the other physicists working in Rome were Bernardo Cacciapuoti (1913–1979), Marcello Conversi (1917–1988), Bruno Ferretti (1913–), Ettore Pancini (1915–1981), Oreste Piccioni (1915–2002), and Gian Carlo Wick (1909–1992).

12 Some historical notes on physical studies in Milan before and (mainly) after WWII can be found in: L. Belloni, 'Giovanni Polvani e l'Istituto di Milano', *Il Nuovo Saggiatore* 4 (1988), 35–49; G. Tagliaferri, 'Le scienze esatte all'Università di Milano', *Storia di Milano*, vol. 18 (Roma: Istituto della Enciclopedia Italiana, 1995), 659–77.

13 On the history of physics in Turin, see: V. De Alfaro, 'Fisica' and C. S. Roero (eds), *La Facoltà di Scienze Matematiche, Fisiche e Naturali dell'Università di Torino 1948–1980*, vol. 1 (Torino: Deputazione Subalpina di Storia Patria, 1999), 207–80.

14 [...] você tem que escolher se a sua escolha é para o presente ou para o futuro. Se é para o presente, a coisa é clara: você entra para o exército de qualquer tipo, oficial ou clandestino, mas entra. Se não, você decide que a melhor coisa é escrever um livro, propondo a si mesmo que este livro vai substituir sua ação e algo vai ficar para preparar o futuro dos jóvens. WAIO: 6.

15 WAIO: 14.

16 On Occhialini's activity in Brazil, see: L. Gariboldi, 'Occhialini's scientific production between the two English periods', SLBO, 71–77; A. M. Ribeiro de Andrade, 'Occhialini's trajectory in Latin America', SLBO, 51–69.

17 On Damy de Souza Santos, see: CBD, 517–30; J. Schober and R. Belisário, 'Marcelo Damy de Souza Santos', *Ciência e Cultura*, 55 (1997), 73–129.

18 On Mário Schönberg (1914–1990), see: CBD, 89–101.

19 Occhialini's main scientific papers during the Brazilian period are: G. Occhialini, 'A simple type of non-ohmic resistance for use with Geiger-Müller counters', *Journal of Scientific Instruments*, 15 (1938), 97–99; G. Occhialini and M. Schönberg, 'Sobre uma componente ultra molle da radiação cósmica', *Anais da Academia Brasileira de Ciências*, 11 (1939), 351–55; 12 (1940), 197–202; G. Occhialini, 'Contributo allo studio dell'effetto di latitudine per gli sciami', *Anais da Academia Brasileira de Ciências*, 12 (1940), 39–44; G. Occhialini, 'Sur la radioactivité béta du rubidium', *Anais da Academia Brasileira de Ciências*, 12 (1940), 155–8; G. Occhialini and M. Damy de Souza Santos, 'Effetto dell'eclissi totale di sole del 1° ottobre sull'intensità della radiazione cosmica', *La Ricerca Scientifica*, (1940), 792; G. Occhialini, P. Pompéia and J. Saboya, 'Nota sobre a estabilização de tensão em corrente alternada', *Anais da Academia Brasileira de Ciências*, 12 (1940), 349–52; G. Occhialini, 'Contributo allo studio della componente ultramolle della radiazione cosmica', *La Ricerca Scientifica*, (1941), 1193–5; G. Occhialini and M. Damy de Souza Santos, 'On a Method of Recording Random Events', *Anais da Academia Brasileira de Ciências*, 13 (1941), 57–62; G. Occhialini, 'Metodo per la stabilizzazione di alte tensioni', *La Ricerca Scientifica*, (1942), 319–21.

20 WA c02 p09 2.354.

21 WA c17 p59.

22 Wataghin não permitia discussão política. A gente vivia num mundo puramente científico. A única preocupação era a ciência; CBD: p. 520.

23 Sair completamente de São Paulo era a única coisa que tinha a fazer *para deixar as pessoas livres para trabalhar sem suspeita.* [...] Eu visualizei a possibilidade de que, em certo momento, uma suspeita, uma sabotagem no laboratório, poderia jogar sobre eles a responsabilidade através de mim; WAIO: 40. Evidence is mine.

24 On Carlos Chagas Filho, see: CBD, 55–63.

25 WAIO: 44.

26 Occhialini worked at the Cavendish Laboratory with Patrick Maynard Stuart Blackett (1897–1974) in 1931–1934. They together built the controlled cloud chamber and discovered the pair production. On their work at the Cavendish, see: P. M. S. Blackett, 'The Old Days of the Cavendish', *Rivista del Nuovo Cimento*, numero speciale (1969), xxxii–xxxix. On the discovery of the positron and the 're-discovery' of the same by Blackett and Occhialini, see: M. De Maria and A. Russo, 'The Discovery of the Positron', *Rivista di Storia della Scienza*, 2 (1985), 237–86; N. R. Hanson, *The Concept of the Positron* (Cambridge: Cambridge University Press, 1963); X. Roqué, 'The Manufacture of the Positron', *Studies in History and Philosophy of Modern Physics*, 28 (1997), 73–129. On Blackett's activity during WWII, see: M. J. Nye, *Blackett. Physics, War, and Politics in the Twentieth Century* (Cambridge MA: Harvard University Press, 2004).
27 Blackett to Foreign Office, 23 February 1943. BP J-59.
28 British Embassy in Rio de Janeiro to Foreign Office, 18 May 1943. BP J-59. Emphasis is mine.
29 Foreign Office to Blackett, 27 October 1943. BP J-59.
30 Não obstante não tivesse tido contatos, sabia que a bomba atômica era possível. Pensava que os alemães a estivessem construíndo e esperava que, do outro lado, também; WAIO: 42.
31 Blackett to Foreign Office, 1 November 1943. BP J-59.
32 Blackett to Foreign Office, 30 October 1943. BP J-59. Emphasis is mine.
33 Blackett to War Office, 5 November 1943. BP J-59.
34 War Office to Blackett, 4 November 1943. BP J-59. Emphasis is mine.
35 War Office to Blackett, 4 November 1943. BP J-59.
36 Wataghin to Blackett, 14 October 1944. BP J-59.
37 Wang to Occhialini, 25 July 1944. WA c02 p06 2.102.
38 Mendes da Rocha to Damy de Souza Santos, 15 September 1944. WA c01 p03 1.175. Pegado to Damy de Souza Santos, 2 October 1944. WA c01 p03 1.178. Unknown sender to Chagas, 6 October 1944. WA c02 p06 2.106.
39 See also the programme of the course of Advanced Physics, WA c05 p23 1.1.12.
40 Registration Certificate issued on 23 January 1945. ODA 4 1.1.
41 Quando procurei os ingleses eu queria ser paraquedista e comecei a treinar para isso. Não acreditava jamais na minha possibilidade de voltar a fazer física ou qualquer outra coisa. Era praticamente uma espécie de suicídio. Só quando fui para a Inglaterra descobrí que os ingleses não estavam interessados em mim como paraquedista ou para trabalho de guerra [...]. WAIO: 46.
42 Department of Scientific and Industrial Research to Ministry of Labour and National Service, 31 January 1945. BP J-59. Emphasis is mine.
43 Bragg to Blackett, 10 February 1945. BP J-59. Emphasis is mine.
44 General Electric Company to Blackett, 16 March 1945. BP J-59. General Electric Company to D.S.R.E., 17 March 1945. BP J-59.
45 Unknown sender to Paterson, 28 March 1945. BP J-59.
46 Aliens War Service Department to General Electric Company, 30 March 1945. BP J-59.
47 On the discovery of the pion by the group led by Cecil Powell, see: L. Gariboldi and O. W. Lock, 'Occhialini's contribution to the discovery of the pion', SLBO, 79–92; O. W. Lock, 'Origins and early days of the Bristol school of cosmic-ray physics', *European Journal of Physics*, 11 (1990), 193–202; A. M. Ribeiro de Andrade, 'The socio-historical construction of the discovery of the π-meson', H. Kragh, G. Vanpaemel and P. Marage (eds) *History of Modern Physics*, (Turnhout: Brepols, 2002), 313–21.
48 A visão que procurei dar é sobre no que diz respeito a escolha: entre fazer pesquisas ou fazer guerra, entre fazer o terrorista ou fazer o cientista, entre lutar contra uma ditadura ou ficar num laboratório. WAIO: 56.
49 Tinha deixado o Brasil, todo mundo me dizia que era besteira; tinha sonhado voltar para a Itália sobre um carro armado, como um conquistador; tinha sonhado em vão fazer trabalho de guerra. *Nada tinha se realizado*. WAIO: 49. Emphasis is mine.

3 Scientific research in the Second World War

The case for Bacinol, Dutch penicillin

Marlene Burns

In November 1945, the recovery of Maria Geene, a patient in Delft's Bethel Hospital, signalled the success of the development of penicillin at NV Nederlandsche Gist- en Spiritusfabriek (NG&SF: Netherlands Yeast and Spirits Factory) in Delft, The Netherlands, during the Second World War, whilst under occupation by Nazi Germany.

Yet, in relating the history of the development of penicillin, David Wilson begins by stating what he calls 'the standard version of the penicillin story' thus:

> … penicillin, the first of the antibiotic drugs. First observed by Sir Alexander Fleming in 1928 when he noticed that a stray mould had killed germs on one of his culture plates. Developed by Lord Florey and Professor Sir Ernst Chain at Oxford in 1940. Mass produced by the U.S. pharmaceutical industry, it saved the lives of thousands of Allied servicemen and came into world use after the end of the Second World War.[1]

However, Wilson submits that while no single item of this story is positively untrue, the whole adds up to a 'myth'.[2] Beginning with Fleming, Wilson claims he misinterpreted and misunderstood what he saw on his laboratory plate. He never found what was causing the effect he saw and never showed that 'penicillin' had any therapeutic or curative effect. Florey and Chain did indeed develop penicillin into a drug in their Oxford laboratory but this was not what they had set out to do in their original research programme. It had been a purely scientific investigation into the phenomenon of bacterial antagonism. Moreover, although mass production did take place in the United States, large-scale production also took place in Britain and Canada. However, for Wilson, the greatest distortion of the truth comes simply in the presentation of the penicillin story in a chronological order. While scientists, and most of the rest of us, have been brought up to believe that there is a steady build-up of knowledge and experimentation from the first observation of biological activity until the final product is marketed, this was certainly not the process in the case of penicillin. As Wilson points out, throughout its story penicillin is marked by the effects of luck, both good and bad, and sheer chance.[3]

Yet, while Wilson continues by relaying the history of penicillin, his sources stay rooted in the British/American 'myth'. He makes no mention of the fact that mass production of Dutch penicillin started at NG&SF just after the Second World War. Nor that Gist Brocades, as NG&SF had become, had gone on to be one of the world's largest producers of bulk penicillin by the mid-1960s.[4] Nor that Gist Brocades continued penicillin production as part of Dutch State Mines (DSM) during the 1990s. Little, therefore, is known of the original Dutch company, which at the end of the Second World War was poised to enter the penicillin world through research that had taken place during the war years. It is the remit of this publication to bring this part of the history of the development of penicillin to the fore.

To begin with, the background to this history remains within the experience of The Netherlands during the Second World War. Following Hitler's *blitzkrieg* tactics of 10 May 1940, the Dutch Queen, Wilhelmina, and most of her Cabinet were forced to flee the country. They became a Government-in-Exile in London and, for the following five years, The Netherlands remained under Nazi occupation. In effect this meant that the whole country was cut off from the outside world.

From the history of penicillin perspective, chronologically, the occupation of The Netherlands was approximately just three months before the first publication of Florey and his Oxford team of August 1940. The second publication from Oxford came in August 1941 – a full year after occupation. Added to this, following the success of the 'wonder drug' penicillin in the treatment of battlefield wounds sustained during the 1943 North Africa campaign, from late 1943 to early 1946 there was an Allied embargo on publications concerning the production and chemistry of penicillin. No information on the development of penicillin was allowed outside Britain and the USA.

How then, during a time of embargo, could those at NG&SF hear of the 'wonder drug' penicillin? How was the development of penicillin at NG&SF possible? The company was, after all, as its name suggests, a provider of yeast and spirits. Who were the researchers? Moreover, given the constraints of occupation, how was this research kept secret? What role, if any, did the 'luck' and 'sheer chance' in Wilson's penicillin 'myth' play in the development of Dutch penicillin? In the following, the author will address these questions.

NG&SF: The Pre-war Years

Since its foundation under Jacques van Marken in 1869, NG&SF had built up a prestigious reputation in the fermentation world. It was the producer of a market leader in the yeast industry with *Koningsgist*.[5] By 1906, François Gerard Waller (Waller Sr), van Marken's nephew, was in charge at Delft and initiated a series of expansion. New production facilities were established in Bruges, Belgium; London and Manchester, Great Britain; Lisbon, Portugal; and, Monheim, Germany. During the First World War, shortages of raw materials led NG&SF to adopt the Danish 'fed-batch' process whereby all carbon sources were utilised to make yeast. This had the advantage that almost no alcohol, the usual offshoot of the fermentation process, was formed and the company was able to concentrate on fermentation.[6]

In 1923, the son of Waller Sr, François Gerard Waller (Waller Jr), joined the company. A trained chemical engineer, his academic training came from Delft's Technische Hoogeschool (TH),[7] and, in particular, from the microbiology laboratory of Professor Albert Jan Kluyver. Waller Jr was known for his keen interest in research and development. By the late 1920s, production processes at NG&SF had been developed for butanol and acetone for the paint industry, and ether was chemically produced. In the yeast industry, NG&SF cornered the market with the production of Engedura, dried yeast. There is no doubt that it was Waller Jr's passion for research and development that encouraged these diversifications.

Between 1928 and 1933 NG&SF's research department was strengthened by the recruitment of three young microbiologists/biochemists, A. P. Struyk, A. A. Stheeman and B. Elema, all of whom were graduates from Kluyver's TH laboratory. In 1933, Waller Jr became Director of Research and Production and by the late 1930s the company headquarters in Delft had laboratories for research and development, a library, instrument makers, glassblowers and an extensive and well-trained staff of biochemists and microbiologists.[8] They were accepted authorities in their field.

NG&SF under Occupation

Under occupation, NG&SF's prestigious reputation in the fermentation industry allowed them to stay in charge of the factory's routine. They were the producers of a necessity for a dietary staple – bread. As a result, NG&SF employees received 'required worker' status and, fermentation being a round-the-clock process, these protected workers could come and go, even during curfew.[9] Although it must be said that all of their homes were to be found in the 'Agneta Park', an area of company-linked housing bordering the factory grounds.[10]

As the war progressed, however, overall production at NG&SF was reduced. Cut off from their affiliates in Belgium, Germany, the UK and Portugal, their yeast products became restricted to the requirements of the local market. Added to which, in 1943, Albert Speer classed alcohol as a luxury and production was cut. Raw materials, too, which could be obtained only through the Dutch *Rijksbureaus* (State Departments), were stringently rationed.[11]

Some NG&SF workers were deported to work in Germany, but, it is said, that NG&SF management took pride in the fact that where possible, placements were found for these workers in their daughter company at Monheim. It was felt that there they might receive reasonable treatment. NG&SF management also sought solutions to the increasing food shortages their workers faced. Subsidised 'factory meals' were introduced. Made in the company's central kitchen, the cost was '5 cents a plate'.[12] While these meals offered a welcome supplement for NG&SF families they further reflect the increasing difficulties of the time.

At this time, too, research was undertaken at the behest of the Dutch organisation for Nutrition and Food, a section of TNO,[13] in the production of vitamin C. This was done in collaboration with Shell and the Chemische Fabriek Naarden. Vitamin shortages further provided the incentive to explore yeast-derived vitamins such as vitamin B_1. NG&SF research workers also produced food supplements for the general population with the development of food enhancers, namely Gistex and Aromex.[14]

At the beginning of occupation, therefore, the experience of NG&SF appears relatively good. Management stayed in place and workers received 'required' status. However, as the war went on, the commercial situation worsened due to lack of materials, and personnel became compromised. To survive, NG&SF needed to fill its fermenters. Ultimately, however, this lack of materials stimulated 'new research' and with this new research came the development of new technical skills and processes. New processes which Waller Jr submitted: 'would stand us in good stead in the production of penicillin'.[15]

The Idea

> They were exciting days for us and not only because of the advance of the Allies.[16]

In an interview in the company newspaper, *De Fabrieksbode*, given to celebrate his 65th birthday, Waller Jr is quoted as relating that he first heard of the new Allied 'wonder drug' from listening, clandestinely, to the radio and from propaganda material. Until recently, it had been thought that the radio programmes

Figure 3.1 F. G. Waller Jr.

referred to were programmes transmitted either through *Radio Oranje* (Radio Orange), the Dutch sender based in London, or the BBC.

However, no reference to penicillin in Radio Orange broadcasts has yet been found. Also, the BBC Archives contain no 'news bulletins' of the time. The BBC Archive does, though, contain transcripts of several programmes relating to the development of penicillin and transmitted by the BBC from 1943 onwards.[17] Nonetheless, none was broadcast on a wavelength pointed to listeners in The Netherlands. It is difficult to ascertain, therefore, the source of radio information gleaned by those in Delft.

Be that as it may, it has been broadly accepted in the past that the propaganda material referred to by Waller Jr was the *Vliegende Hollander* (Flying Dutchman). Yet research with the *Vliegende Hollander* has uncovered no mention of penicillin. There are, however, two articles on penicillin in another propaganda publication, namely *De Wervelwind* (The Whirlwind). Unlike the weekly *Vliegende Hollander*, *De Wervelwind* was a monthly 'magazine', and the articles on penicillin are contained in the issues of December 1943 and April 1944. However, records show that the December 1943 issue was *NV – niet verspreid* (not dropped) and that the April 1944 issue was dropped over 'Appingedam, Delfzijl, Hilversum, Lage Vuursche, Keulen, Rucphen', all some considerable distance from Delft. At the same time, the timing of the April 1944 *Wervelwind* as deliverer of information on the Allied use of penicillin is, as will be shown, well outside the timing of NG&SF's first research with *Penicillium* moulds. The question of how information about penicillin initially reached Waller Jr and his research workers in Delft, therefore, remains debatable.

Starting Position

Nonetheless, the 'starting position' of those at Delft regarding their laboratory research with *Penicillium* strains remains uncontested. As shown earlier, the first publication of Florey and his Oxford team in August 1940 had not taken place until after the occupation of The Netherlands in May 1940. Since occupation, The Netherlands had been cut adrift from research taking place in the Allied world. When news of penicillin development became a fact in Delft, the only scientific information available to NG&SF researchers would have been the available pre-war publications.

To an extent, the first NG&SF R&D Report on the search for an antibacterial substance confirms this. Written by A. P. Struyk and dated March–June 1944, R&D Report 412 clearly sets out Struyk's background materials.

Although here it has to be noted that Struyk's original report refers only to journals and page numbers, research in *Chemical Abstracts* has produced specific article titles. These are:

Fleming, A., 'On the Antibacterial Action of Cultures of a *Penicillium* with Special Reference to Their Use in the Isolation of *B. influenzae*', *British Journal of Experimental Pathology*, 10, (1929), 226–36.

Clutterbuck, P. W., Lovell, R. and Raistrick, H., 'The Formation from Glucose by Members of the *Penicillium chrysogenum* Series of a Pigment, an Alkali-Soluble Protein and Penicillin – the Antibacterial Substance of Fleming', *Biochemical Journal*, 26, (1932), 1907–18.

Waksman, S. A., 'Antagonistic Interrelationships among Microorganisms', *Chronica Botanica*, 6, (30 December 1940), 145–8.

In the UK, Fleming's original article had been followed by Clutterbuck's research but both had been inconclusive on how to produce purified penicillin in quantity. In the United States, Selman Waksman, too, had published extensively on 'antagonistic relationships' in the pre-war years but not specifically on penicillin as an antibacterial substance.[18] All three publications would, however, have been available as pre-war academic literature.

Yet the above publications make up only half of Struyk's sources, and it is from those remaining in Struyk's list that we can begin to glean the existence of a dissemination of information concerning penicillin during the war years. The three other sources are:

Vonkennel, J., Kimmig, J. und Lembke, A., 'Die Mycoine, eine Neue Gruppe Therapeutisch Wirksamer Substanzen aus Pilzen', (The Mycoins, a New Group of Therapeutically Active Substances from Fungi), *Klinische Wochenschrift*, 22, 16–17, (17 April 1943), 321.

Kiese, M., 'Chemotherapie mit Antibakteriellen Stoffen aus Niederen Pilzen und Bakterien', (Chemotherapy with Antibacterial Substances from Moulds and Bacteria), *Klinische Wochenschrift*, 22, 32–33, (7 August 1943), 505–11.

Penau, H., Levaditi, C., et Hagemann, G., 'Essais d'Extraction d'une Substance Bactericide d'Origine Fungique', (Attempts to Extract a Bacterial Substance of Fungal Origin), *Bulletin de la Societe de Chimie Biologique*, 25, (1943), 406–10.

Chronologically portrayed, all of the above were published during the year 1943. All illustrate a widening circle of research into mould-based, penicillin-like, antibacterial substances. They also illustrate the wartime dissemination of information on penicillin in Germany and occupied France. As such these publications act as a marker to the Allied 'need' for an embargo on information over the development of penicillin. In particular, M. Kiese of the University of Berlin listed in his publication a very impressive sixty-one references on penicillin and antibacterial substances that had been published between 1923 and 1943. His publication also covered the research at Oxford by Florey et al. in detail.

But Struyk's sources raise yet another question: How did Struyk, a microbiologist at a yeast factory in occupied Delft, obtain these foreign academic publications? Again, while Struyk's report does not list the source of his material, evidence pointing to his source has been found in the Technical University of Delft where the Kluyver Archive contains photocopies of all of Struyk's source material. All bear the stamp *Bibliotheek D.B.M.* (Library Delft, Bruges, Monheim). It would appear, therefore, that the basis for Struyk's research came innocently enough through NG&SF's own inter-library loan system.

Strain Selection

Struyk's Report 412, March–June 1944, further shows the manner of his strain selection. According to Struyk, he received from the Centraalbureau voor Schimmelcultures (CBS: National Collection of Fungal Cultures) twenty-one fungal strains. These consisted of eighteen *Penicillium* strains and three *Aspergillus*. To this Struyk added two fungal strains that had been found on old cacao powder.

Research in the archive of the CBS has shown, however, that the then director of CBS, Professor Johanna Westerdijk, did not supply Struyk's twenty-one strains en bloc. She did so on a strain-by-strain basis. Also, the contact Westerdijk had with NG&SF was not through Struyk or Waller Jr but with another NG&SF Yeast Division staff member, namely Johannes Rombouts.

Correspondence between Westerdijk and Rombouts began on 19 January 1944 with the delivery from CBS to NG&SF of twelve moulds. On 24 February 1944, Rombouts thanked Westerdijk for her offer to supply NG&SF with *Penicillium* strains free but reported that the directors preferred at this difficult time to give CBS financial support. He also asked that should she hear of other moulds producing a 'good bacteriostatic substance' he would be pleased to receive them.[19] More strains followed on 15, 16 and 21 March; 1 April; and 15 and 24 May 1944.[20] Struyk's research, therefore, was not limited to one experiment as his report might imply. From Rombouts' contact with Westerdijk it appears that Struyk's research was more an ongoing, continuous process.

Strain Evaluation

Report 412 illustrates the methodology Struyk followed to evaluate the strength of any antibacterial produced by his twenty-one *Penicillium* strains. Using an agar block test and *Micrococcus aureus (Rosenbach)*, an old name for *Staphylococcus aureus*, which he had obtained from Kluyver's laboratory, he developed a 'zone of inhibition'. From this 'zone' the activity of the strains could be compared. To monitor differences he created a 'Delft Unit' with which to define any antibacterial activity. By so doing, Struyk's basic aim was to reproduce Fleming's initial findings.

Struyk's results show that seven of his experimental strains produced an antibacterial substance. Ultimately, the mould culture with the highest yield and the one chosen for further study was sixth on Struyk's list, P6: *Penicillium baculatum* Westling. Struyk named this substance Bacinol.

Further research with Bacinol is noted in Struyk's following reports, numbered 413 and 414. These R&D reports also reflect research with Bacinol during the period March–June 1944. For example, Report 413 illustrates that if *Penicillium baculatum* was allowed to grow on NG&SF's own Liquitex base for five days at a constant temperature of 26°C and shaken once a day, the results appeared to be identical to those reported by Fleming using a bouillon mash and *Penicillium notatum*. Also, the substance produced by P6 was soluble in acetone and alcohol, which facilitated extraction from the growth mash, and, when mixed with water, its properties were resistant to boiling. Bacinol, therefore, had the same antibacterial and physical properties as penicillin – as described by Fleming.

In all, Struyk's R&D Reports 412, 413 and 414 total twenty-eight pages. They cover the growth, extraction and sterile conditions necessary for the production of the antibacterial substance produced by *Penicillium baculatum*. All three are clipped together to make one document. It is not marked 'secret' and it would appear to have been circulated through normal office channels. The office stamp shows that it was ready for circulation on 29 July 1944 and the intended recipients are noted as F. G. Waller Jr, A. A. Stheeman and J. R. Rombouts.

Serendipity – The Importance of 'Chance'

As stated earlier, David Wilson points to the story of penicillin as marked by the effects of luck, both good and bad, and sheer chance.[21] For the development of penicillin in The Netherlands, the same holds true as illustrated by the experience of one of NG&SF's advisors, Andries Querido.

On 4 September 1944, Querido, a Jewish doctor interned in Westerbork Camp, was transported to Theresienstadt in Czechoslovakia.[22] Before this, however, as a part-time advisor to NG&SF he still had 'Required Worker' status and was permitted to visit the Delft factory, albeit on an irregular basis. It was on what was to be his last visit to Delft that Querido met a fellow Jewish doctor, S. van Creveld, in Amsterdam's Central Station. Van Creveld, Professor of Paediatrics in Amsterdam, was at the time still working. Bursting with excitement, he told Querido that he had just had a visit from a colleague from neutral Portugal. That colleague had brought with him a copy of the latest *Schweizerische Medizinische Wochenschrift* (*Swiss Medical Journal*). Dated June 1944 the whole publication was given over to the subject of penicillin. Querido knew how important this would be to those at Delft and asked if he could borrow it. He assured van Creveld that the journal would be returned to him. Van Creveld agreed.[23]

Critically for those at Delft, this issue of the Swiss Medical Journal contained an article by A. Wettstein. Simply entitled 'Penicillin', it clearly showed the results the Allied had achieved in the development and production of penicillin. For example, Wettstein gave details of penicillin growth on a maize extract; of the scale-up of penicillin production in bottles and porcelain containers; the measurement of strength by the Oxford unit; a dilution method; physical and chemical properties; human studies; animal studies; and named bacteria known to be sensitive or insensitive to penicillin. For the Delft researchers this information proved invaluable. Not only did it confirm their own antibacterial research, it reinforced their commitment to continue.

Yet this issue of the *Swiss Medical Journal* also served to illustrate the ongoing dissemination of information over penicillin during the war years. Written only a year after Kiese's German publication, which had been based on sixty-one sources of references, astonishingly Wettstein, in neutral Switzerland, could cite 159 sources. At a time of embargo, the jump is enormous.

At Delft, the entire *Swiss Medical Journal* of 10 June 1944 was photocopied and circulated at least thirteen times. The Kluyver Archive holds Kluyver's copy. The protective cover is stamped '*Bibliotheek D.B.M*' and the label reads '*Photocopie nr. 13 in 4 voud*' (Photocopy number 13 in 4 fold). Under occupation, therefore, the copies were made, quite simply, in NG&SF's library and laboratory research continued.

Increasing Production

As research continued, Rombouts showed that Bacinol could be grown on the surface of a liquid medium in Jena and Roux flasks, and Stheeman investigated a quicker method of extracting the harvested culture fluid using ether in place of alcohol.

To further enhance the growth of Bacinol, Struyk tried various types of flat glass and enamel containers. In the end, he chose what is known as the 'natural fermenter', milk bottles.[24]

Figure 3.2 Milk bottles.

NG&SF employee Klaas Scheurkogel describes the process. The milk bottles were kept for 10–12 days at a temperature of 25°C. After processing the fluid produced, the result was fairly crude penicillin. Sometimes the surface culture became contaminated, which made the content of the bottle unusable. This had to be disposed of and the process started again. From time to time such 'calamities' seemed insurmountable but the Delft researchers kept going. By August 1944 they had a small amount of a gold-brown substance, which, according to Scheurkogel, gave 'all the desired properties'.[25]

Research – A Temporary Stop

Scheurkogel continues that at this point research into the development of Bacinol at NG&SF was scaled down. While it was generally felt that the war would soon end there was a fear in Delft that the *Wehrmacht* might still be able to profit from their antibacterial substance. These two factors led to the decision to put a temporary stop to NG&SF's research.

Waller Jr takes the reasoning for the halt in the development of Bacinol further with the statement that:

> By around Dolle Dinsdag we had a small amount of a substance which we hoped, and later to our joy proved to be penicillin.[26]

By pinpointing 'Dolle Dinsdag' (Mad Tuesday) Waller Jr points to 5 September 1944. By this time France and Belgium had been liberated. As the Allies moved towards The Netherlands it was confidently expected that liberation would follow. On 5 September the BBC erroneously reported the liberation of Breda and the Dutch population lined the streets in order to greet the Allied forces. This did not happen. Operation Market Garden failed at Arnhem. The south of the country was liberated but the north and western provinces remained occupied. Ultimately, for those caught within the occupied provinces, there remained the devastation of the *hongerwinter* (Hunger Winter).

The 'Hunger Winter' did not mean that there was no food. Little though it was, some food was available. In retaliation for the Dutch workers railway strike, which was an attempt to help the Allied cause at Arnhem, the occupier refused the transport of food supplies to the western provinces. In the face of one of the coldest and bitterest of winters, the Dutch population was left to flounder.

Re-establishing Research

Scheurkogel claims that it was during the dark days of the winter of 1944–1945 that the decision was taken to re-commence NG&SF penicillin research. This led to the further purification of the end product and by the end of April 1945 a few ampoules of Bacinol had been created.[27]

NG&SF's Research and Development Archive confirms this. For example, Rombouts, with his assistant Ans Addeson, went on to test for toxicity in

Staphylococcus aureus-infected rabbits and mice; between July 1944 and March 1945 R&D Reports 847–904 show that Stheeman, with his assistants Knotnerus and Mathu, continued to improve penicillin extraction methods from the broth culture; and, in Report 243, which covers April–May 1945, Stheeman signalled the differing levels of success in the search for an improved 'mash' with which to 'feed' Bacinol by growing *Penicillium baculatum* on sugars, beet pulps and grain mixes. His conclusion was that the most successful was quite simply 'grain mash'.

Penicillin and Liberation

Officially, it was to be 5 May 1945 before liberation came to the people of the western Netherlands. Before this, however, an agreement was reached between the occupier and the Allies, which allowed the dropping of food by British and American bombers to the beleaguered Dutch. The British started these drops on 28 April 1945 at the airfields of Ypenburg (Delft), Duindigt (The Hague), Valkenburg (Leiden) and Waalhaven (Rotterdam). Wider drops came with American help. The occupier had agreed to stand aside, allowing the distribution of the food to be undertaken by the Dutch themselves.

Until now, most sources relating the history of NG&SF penicillin claim that American penicillin was included in the food dropped at Ypenburg, Delft. Some say that a Delft doctor, E. Verschuyl, took 'dropped' penicillin to NG&SF against which they compared their own antibacterial substance. Others simply report that American penicillin was part of the food drop.[28]

It is difficult to see why the Allies would have dropped penicillin. At the time, penicillin was restricted to military use only. There was no surplus. Added to that, Dutch doctors had no experience with penicillin as a medical treatment. They did not know its properties or how to use it. At the time, penicillin could only be administered by intramuscular or intravenous injection. In powdered form it had to be mixed with sterile water before an injection took place. Injections had to be sustained on a twenty-four-hour basis. How, therefore, could it have been possible for the new 'wonder drug' to be part of any food drops, let alone only at Delft?

Bacinol: Dutch Penicillin

Nevertheless, at the end of the war American penicillin did reach Delft. On 15 May 1945, Stheeman, with his assistant C. W. F. Spiers, wrote in R&D Report 750 of the analysis of a penicillin preparation. While they do not state the source of the penicillin preparation used, their conclusion was that it did not have the potency stated by the manufacturers. Whether this was because it had been kept too long or at too high a temperature they could not say. What they did say was that the preparation they had analysed was not pure. There was, therefore, not point in using it to determine the properties of penicillin.

However, research at NG&SF did not stop. For example, Stheeman's R&D Reports 244–246 covering the months June–July 1945 shows the analysis of a sample of American penicillin made by Chas Pfizer & Co and supplied by Upjohn of Kalamazoo. He also indicated the results of experimentation with different mash nutrients as he sought to find a higher-yielding mould strain, one that would surpass the level of penicillin production then achieved by *Penicillium baculatum*. In July 1945, therefore, barely two months from liberation the Delft Team knew they were in possession of an antibacterial substance similar to the penicillin mass-produced in the United States during the war years.

First Clinical Application

In November 1945, the recovery of Maria Geene, a patient in Delft's Bethel Hospital, signalled the success of the development of NG&SF's Dutch penicillin. However, research shows that she was not the only patient to receive treatment at that time. As will be shown, another young woman, who, like Maria Geene, was dying of septicaemia, received treatment with Bacinol.

The temperature charts below illustrate their dramatic recovery. The first is of Maria Geene. At the time aged twenty-one, she was admitted to the Bethel hospital on 15 November 1945. She was critically ill with a staphylococcus infection. Her temperature was 39–40°C. She received an injection of 50,000 units of Bacinol. Her temperature returned to normal and she was discharged on 29 November 1945.[29] In total, her admittance to discharge had taken only 14 days.

In her 1991 television interview with Willy Lindwer, she described her feelings of joy when she realised her life had been saved. The medical staff, too, were overjoyed and showed their happiness and relief as they hugged her.[30]

The second patient remains unknown. Then, she was eighteen years old. She was admitted to the Bethel on 26 November 1945. Her temperature was 40–41°C. On that day she received 50,000 units of Bacinol. Intravenous injections continued on 27 and 28 November with 100,000 and 150,000 units, respectively. Her infection cleared and her temperature returned to normal. She was discharged 14 December 1945. Her complete recovery had taken just nineteen days.

NG&SF Penicillin: The Position 1945–1946

In 1945, the US and the UK were producing as much penicillin as possible but it was not enough to fill the demand. In addition, American penicillin, when produced in other countries, came under licence. At the end of the war, the position of those at NG&SF was that they knew they could produce penicillin equal in quality to that of the American and British companies. There was no need to wait for American licence; they had their own strain, *Penicillium baculatum*. To produce Bacinol, NG&SF had used their own research and development team, and they had developed their own production techniques. Waller Jr's passion for research had instilled 'a will to succeed'.[31] In August 1946, NG&SF penicillin produced on a factory scale came on to the Dutch market. Before this could happen, though, more immediate developments in the large-scale production of Bacinol had to be addressed.

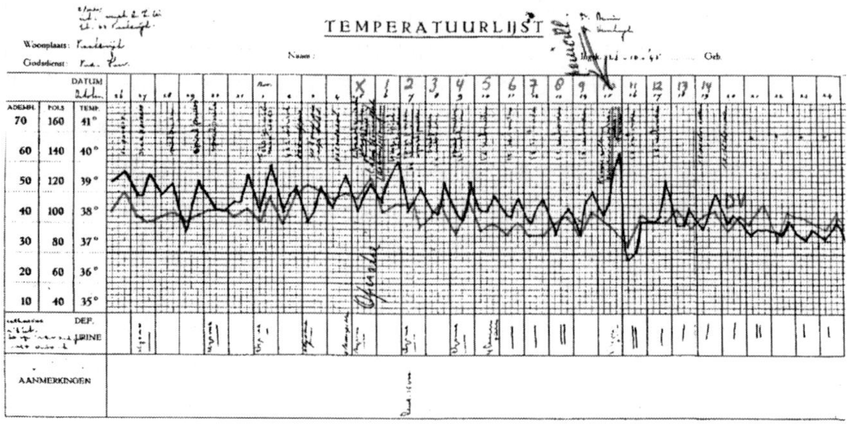

Temperatuurlijst (ziektegeschiedenis I) 26.10.1945 - 05.12.1945. Op 15 november 1945 wordt penicilline intraveneus gegeven.

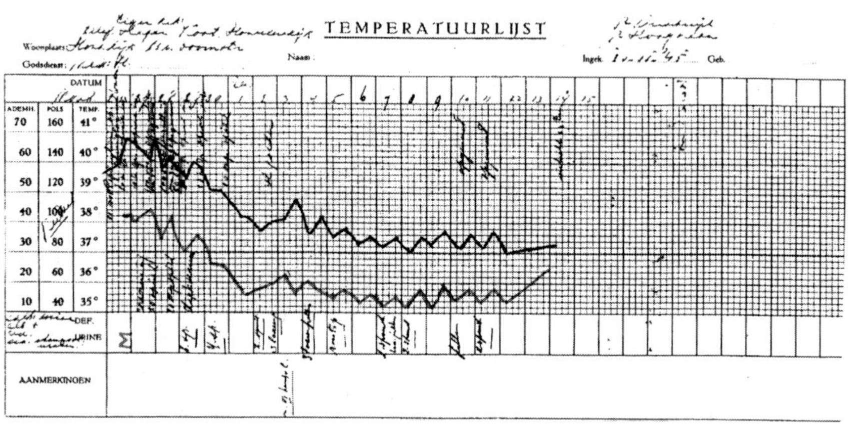

Temperatuurlijst (ziektegeschiedenis II) 26.11.1945 - 14.12.1945. Op 26, 27 en 28 november 1945 wordt penicilline intraveneus gegeven.

Figure 3.3 Temperature charts.

For this, the Delft Team went back to their advisor and mentor, Albert Jan Kluyver. Kluyver had been Professor of Microbiology at the TH in Delft from 1922. With the exception of Rombouts, he had taught them. As stated earlier, Struyk, Stheeman and Waller Jr had graduated under his guidance, following which they remained close friends. Kluyver's position as an NG&SF advisor had been formalised in 1933 and he regularly joined the Monday morning meetings of NG&SF's research laboratory. The influence of Kluyver in the scaling-up of Bacinol from a laboratory bench to an industrial level is plain to see. During the 1930s Kluyver had published on a new concept, deep fermentation technology.[32] Just as the concept of deep fermentation technology had increased the production of penicillin in the US and in Britain, so too did it influence his students at NG&SF.

Factory production of penicillin at NG&SF was successfully started on 15 May 1946 when the first industrial fermentation took place in a 1.5 hectolitre *Ensinkketel* (Ensink tank). Upscaling to 15, 60 and 300 hectolitre fermentation tanks soon followed this,[33] an achievement that can only be described as an incredible rate of expansion.

In order to do so, a new group of workers was established. Johan van den Berg, a member of the first large-scale fermentation team, describes them as 'a group of young men employed purely for the large-scale production of NG&SF penicillin'. They quickly acquired the title 'Penicillin Experts'.[34] H. de Horn, another of the first NG&SF employees to be included in the scale-up of NG&SF's penicillin, paints the scene: 'We had to think on our feet, to solve problems as we went along. If you saw a couple of colleagues deep in conversation, you knew it was about penicillin'.[35]

The team learned freeze-drying techniques from the Blood Transfusion Service in Amsterdam. They also learned how to deal with the sensitivity of penicillin to impurities. For example, the first attempt to protect the tank from impurities was met by sterilising the steel wool filters. However, after the second use, these melted. Eventually, a method was developed called 'double steam sealing' whereby the contents of the tanks had to go through not just one but two steam-filled 'dips' in the extraction pipe. The theory behind this was that impurities might be able to filter through one steam 'dip' but not a second. De Horn, too, developed a system called 'the Hornex', a counter-current device to concentrate the penicillin contained in the slurry. Finally, at a time when the whole of The Netherlands was in need of reconstruction, Waller Jr's desire to increase NG&SF's production of penicillin came with the purchase of the Leidse Machine Fabriek in 1947, which he renamed Leidsche Apparaten Fabriek (LAF). Critically for Waller Jr, the LAF had a fifty-man workforce expert in the production of metal tanks. They were also specialists in stainless steel products. By his purchase, Waller Jr ensured that these talents came to NG&SF. In its first year of production the LAF met 75 per cent of NG&SF's apparatus requirements.[36]

Finally, the Allies lifted their embargo on penicillin data in 1946. But by that time Waller Jr. had taken a critical step at NG&SF. He had moved the Gist- en Spiritusfabriek from being only a yeast fermentation plant into the pharmaceutical world for the mass production of Dutch penicillin.

Antibiotics at Delft

In January 1946, NG&SF established an Antibiotics Department. The first coordinator was Klaas Scheurkogel. Shortly afterwards, R. A. Jellema was appointed head of the first NG&SF Penicillin Department.[37]

Together, their duty was to establish contact with the Dutch medical world and to promote the use of NG&SF penicillin. In order to do so, a Medical Brains Trust was formed. This Trust, chaired by Kluyver, contained a mix of Querido and two respected medical practitioners from Leiden's Academic Hospital, namely Jacob Mulder and Willem Goslings. To this group was added L. E. den

Dooren de Jong, a bacteriologist from Delft's TH, and, as and when necessary, Waller Jr and his staff.

The Medical Brains Trust published *Digesta Antibiotica*, an academic publication given over completely to the 'wonder drug' penicillin. Using the most up-to-date material available from both Britain and the US they wrote articles explaining the manner in which the, by now, various forms of penicillin could be and should be administered. The editorial team – Querido, Mulder and Goslings – also answered questions about penicillin and its use.[38]

Scheurkogel and Jellema had the remit of bringing NG&SF penicillin to the Dutch public. In 1946, penicillin in The Netherlands was rationed as the Government grappled with the cost of import. A voucher system for the distribution of penicillin was introduced and, in August 1946, the Dutch Government allocated the supply of NG&SF penicillin to seven hospitals. These were the Academic Hospitals of Leiden, Utrecht and Groningen; Johannes de Deo, The Hague; Wilhelmina Gasthuis, Amsterdam; St Jacobus Stichting, Wassenaar; and De Gemeente Apotheek, The Hague. As the production of penicillin in NG&SF fermenters increased, so the allocation system increased to meet NG&SF's production capacity.[39]

By the end of 1946, NG&SF was supplying all the penicillin needed by Dutch hospitals. By 1948, NG&SF was supplying all penicillin requirements for the whole of The Netherlands. In 1949, NG&SF started exporting penicillin. The first country to receive it was Indonesia, a former colony. In 1950, NG&SF received the predicate *Koninklijke* (Royal). Fifty years from the end of the Second World War, Gist Brocades, as NG&SF had become, was one of the world's largest producers of bulk penicillin.

In March 2005, DSM, as Gist Brocades had become, closed down most penicillin fermenters at Delft. Almost exactly sixty years from the first large-scale production of penicillin in Delft, 'market forces' have taken the large-scale production of Dutch penicillin to India and China.[40]

Conclusion

Code-named Bacinol, the secret production of penicillin at NG&SF, Delft, whilst under the extreme condition of Nazi occupation, did happen. It ran alongside legitimate wartime research with NG&SF food enhancers Gistex and Aromex and a new joint venture for the development of vitamin C with Shell and *Chemische Fabriek Naarden*.

Initially, the true identity of NG&SF's antibacterial substance remained protected because of its name. The name Bacinol was derived from the mould strain *Penicillium baculatum*. During the war years it was never referred to as 'penicillin'. On the one hand, the researchers could not be sure that the antibacterial substance they had 'grown' was equal to Fleming's penicillin. On the other hand, they wished to ensure that their research stayed secret from their German administrator.

Accordingly, experiments with Bacinol were adapted to suit wartime conditions. The growth of Bacinol took place quite simply on NG&SF's own fermentation mash, Liquitex. Milk bottles provided the container.

Yet, as in all experimental procedures, reports had to be written and, as we have seen, these were clear and concise in methodology and observation. Consequently, while these reports highlight the fact that contemporary scientific journals were in some way reaching those within NG&SF, critically, they reflect a confidence when addressing the then recent microbiological research and an ability to put that research into practice. Ultimately, the fact that research with Bacinol was kept from their German occupier merely adds to the achievement of the NG&SF team.

When considering the success of the wartime secrecy surrounding the laboratory development of Bacinol at NG&SF, this success could be put down to the fact that NG&SF was a 'family' concern. They were accepted experts in their field. They knew their business and how to keep trade secrets. As de Horn points out, 'It was not unusual to see milk bottles on their side growing "stuff". That was fermentation and NG&SF knew more about fermentation than their German occupiers'.[41]

In the end, the development of penicillin was a technical problem that required in-depth knowledge of microbiology, fermentation and recovery. The Delft Team had all of that. NG&SF had unique access to Delft's TH where Professor Kluyver was a world authority in the microbiological field. Many of Kluyver's former students were employed by NG&SF, while Querido's experience brought the influence of 'chance'. All of this was added to the fact that F. G. Waller Jr was a determined, inspirational, leader. He had an excellent technical grasp and a flair for improvisation. The Delft Team was small and cohesive, with no bureaucracy and with short lines of communication. They had a 'will' to succeed.

On the face of it, however, this author is of the opinion that it was simply not expected by the Reich that a yeast factory in Delft could develop something like the 'wonder drug' penicillin.

At the end of the war, NG&SF took the massive step of including fermentation-based pharmaceutical products in its remit. At a time when the whole of The Netherlands required reconstruction this was a step that demanded considerable investment. Here, again, Waller Jr's determination showed through. He invested not only in plant and machinery but also expanded NG&SF personnel and advisors.

In 1968, Koninklijke (Royal) NG&SF merged with the pharmaceutically based company Brocades, Stheeman & Pharmacia to become Gist Brocades. Yet, in March 2005 – sixty years from the start of penicillin production at Delft – DSM, as Gist Brocades had become, stopped the production of most penicillins at Delft. The reason given for this closure was: market forces.

Ultimately, the perspective of this publication has been to record a part of scientific research that took place during the Second World War. But it has also been to bring the extraordinary experience of those at NG&SF into perspective in the history of penicillin.

David Wilson highlights the penicillin 'myth' and the emphasis placed on research in Britain followed by large-scale production in the United States in a more or less chronological order. He also points to the involvement of penicillin production in

Canada. In the history of penicillin, however, little is known or recorded of the Dutch company that was successful in the wartime development of a penicillin-like, antibacterial substance. Little is known of The Netherlands Yeast and Spirits Factory, Nederlandsche Gist- en Spiritusfabriek, which, at the end of the Second World War, after five years of occupation and isolation, stood on the brink of success in the large-scale production of their own penicillin – Bacinol, Dutch penicillin.

The Delft Team:

Leader: F. G. Waller Jr, NG&SF Deputy Director, Delft
Researchers: A. P. Struyk, A. A. Stheeman, J. R. Rombouts
Research Assistants: Lagendijk, Knotnerus, Mathu, Spiers, Addeson
Fermentation: W. A. Verkennis, J. M. Klokgieters
Clinical Application: E. Verschuyl
Upscaling: W. Berends, H. M. de Horn, L. M. Rientsma
Upscaling Assistants: Jongbloed, van den Berg, Elzenga, ter Horst, Kamps, Mensinga, Mostert, Saltet, van der Zijde
Antibiotics Department: K. Scheurkogel and R. A. Jellema
Advisors: W. H. van Leeuwen, NG&SF President; H. F. Waller, NG&SF Deputy Director, brother of Waller Jr; Professors A. J. Kluyver, TH, Delft, and J. Westerdijk, CBS, Baarn; and, Physicians: A. Querido, J. Mulder and W. R. O. Goslings of Leiden University Hospital

Notes

1 D. Wilson, *Penicillin in Perspective* (London: Faber & Faber, 1976), 3.
2 Wilson 1976, 4.
3 Wilson 1976, 5.
4 NG&SF merged with NV Brocades Stheeman & Pharmacia to become NV Gist Brocades in 1968.
5 Literally translated: King's Yeast.
6 M. Burns and P. W. M. van Dijck, 'The Development of the Penicillin Production Process in Delft, The Netherlands, During World War II Under Nazi Occupation', *Advances in Applied Microbiology*, 51, (2002), 185–200, at 186.
7 Technische Hoogeschool literally translates as Technical High School but more equivalent to Polytechnic Colleges of the UK; now TUD, Delft University of Technology.
8 Burns and van Dijck 2002, 186–8.
9 M. Burns, 'The Development of Penicillin in The Netherlands 1940–1950: The Pivotal Role of NV Nederlandsche Gist- en Spiritusfabriek, Delft', PhD (History) Thesis, University of Sheffield, Sheffield, England, UK, September 2005, 111–12.
10 Named after Agneta van Marken, wife of NG&SF founder, J. C. van Marken.
11 Burns 2005, 111–12.
12 NG&SF Company Newspaper *De Fabrieksbode*, 30 August 1941; Burns 2005, 117.
13 TNO: Toegepaste Natuurwetenschappelijk Organisatie, Organisation for Applied Scientific Research.
14 B. Elema, *Opkomst, evolutie en betekenis van research gedurende honderd jaren gistfabriek* (Delft: Koninklijke Nederlandsche Gist- en Spiritusfabriek, 1970), 34; Burns 2005, 112.
15 F. G. Waller Jr, *De Fabrieksbode*, 15 October 1960, 269.

16 Waller Jr 1960, 269.
17 BBC Written Archives Centre, Caversham Park, Reading, Berkshire, UK.
18 S. A. Waksman, 1952 Nobel Prize Winner, for his research with the Gram-negative antibiotic *Streptomycin.*
19 CBS Archive, 1944, Correspondence File, No. 176.
20 CBS Archive, 1944, Correspondence File, Nos 511, 513, 514, 515, 516.
21 Wilson 1976, 5.
22 Before his internment in Westerbork, Querido, his wife and child had been interned in Barneveld Camp in January 1943. On 29 September 1943 the 'Barneveld Jews' were transported en bloc to Westerbork.
23 Personal Communication December 1999; A. Querido, *Andries Querido: de binnenkant van de geneeskunde. Een autobiografie opgetekend in samenwerking met Jacky Bax and Ruud Overdijk* (Amsterdam: Meulenhoff, 1990), 93; Burns 2005, 127.
24 Burns and van Dijck 2002, 196.
25 K. Scheurkogel, 'Technische Bereiding van Penicilline', *Chemisch Weekblad*, 45 (29 January 1949), 69–72.
26 Waller Jr 1960, 269.
27 Scheurkogel 1949, 69–72.
28 Elema 1970; Gist Brocades Company Publications: 30 jaar Nederlandse penicilline, 1973; 35 jaar penicilline, 1978; Van Fleming tot Flemoxin Solutab. Markante momenten in 60 jaar penicilline, 1989; *Brood op de plank: 130 Jaar 'De Gistfabriek' in Delft*, 1999.
29 Burns 2005, 145–6.
30 Video: *De revolutie van het geneesmiddel. 50 jaar penicilline.* Producer: Willy Lindwer (AVA Productions BV, Amstelveen, 1991).
31 H. M. de Horn, personal communication, November/December 1999.
32 A. J. Kluyver und L. H. C. Perquin, 'Zur Methodik der Schimmelstoff-Wechseluntersuchung', *Biochemische Zeitschrift*, 266, (1933), 68–81.
33 J. van den Berg, personal communication, April 2005.
34 J. van den Berg, personal communication, April 2005.
35 H. M. de Horn, personal communication, November/December 1999.
36 Burns 2005, 217–18.
37 Later President of Gist Brocades.
38 Burns 2005, 231–2.
39 Burns 2005, 224–7.
40 Gist Brocades became part of DSM in 1998.
41 H. M. de Horn, personal communication November/December 1999.

4 Preventing theft

The Kamerlingh Onnes Laboratory in wartime

Dirk van Delft

On Friday 30 May 1941, Bart Saris received his doctorate from Leiden University for a thesis on thermal conductivity in liquid helium II (the superfluid form of helium). His official supervisor was the physics professor and laboratory director Wander Johannes de Haas.[1] This was a typical Leiden research project focusing on temperatures just above absolute zero (–273°C), carried out at the world-famous Kamerlingh Onnes Laboratory. The Leiden cryogenic laboratory was founded by Heike Kamerlingh Onnes (1853–1926), who was the first to liquefy helium (1908) and who discovered superconductivity (1911).[2] At the time that the degree was awarded to Saris, The Netherlands had been under German occupation for a year. Articles on the subject had already been published in the Dutch journal *Physica* by Saris, Willem Keesom and Keesom's daughter Annie.[3] The last of these articles, authored by Saris and Keesom, had been submitted on 21 August 1940.[4]

In short, Saris had been forced to wait for his doctoral bull for some time. This was because of the closure of Leiden University on 27 November 1940, a direct response to a demonstrative student strike against anti-Jewish measures taken by the German occupiers. Official educational activities came to a halt, and examinations and doctoral defences no longer took place.[5] For those, like Saris, who were almost finished with their doctoral research, the situation was tremendously frustrating.[6] At the urging of the Board of Governors, the Germans permitted examinations and defences once again starting on 30 April 1941, but reinstated the ban on 20 November of that year.[7]

Because of all this, Jaap Kistemaker, another assistant of Keesom who was also researching properties of supercooled liquid helium, had to wait for his doctorate until 21 November 1945, following the liberation of The Netherlands on 5 May.[8] In any case, there were other reasons that Kistemaker's defence could not have taken place much earlier. While research at the Kamerlingh Onnes Laboratory (including liquid helium production) continued through most of the war, it all came to a stop in the summer of 1944, when the Germans requisitioned a variety of equipment. Kistemaker then had no choice but to abandon his work temporarily.

When the time finally came for Kistemaker's defence, his acting supervisor was not Keesom, who had spent almost the entire war sick at home and had not yet recovered. Keesom had put this time to good use, writing his standard work, *Helium*, which was published in 1942. His co-director, de Haas, did not serve as supervisor either; he had been suspended in June 1945 by Leiden University's Purgation Board pending a thorough investigation of his advisory work in 1943–1944 for Cellastic, a German intelligence service based in Paris (discussed below). It was Hans Kramers, a professor of theoretical physics in Leiden, who awarded the doctoral bull to Kistemaker.

Kistemaker, too, had worked for Cellastic, at the request of de Haas. It was in part because Kramers had put in a good word for him that he was permitted to take his degree in November. This certainly was inspired by the possession of ten metric tons of uranium, which had been in The Netherlands since 1939 (thanks to de Haas). Kramers wanted Kistemaker to move on quickly to Niels Bohr's institute for theoretical physics in Copenhagen, where he could master that field, to fulfil the Dutch ambitions in the domain of nuclear energy.

De Haas and the Yellowcake

Professor de Haas, one of the successors (along with Keesom) to cryogenic pioneer Heike Kamerlingh Onnes as director of the Kamerlingh Onnes Laboratory,[9] was a man with sudden fits of inspiration, driven more by intuition than by analytic or mathematical reasoning.[10] On 13 May 1940, just three days after the German invasion, de Haas asked Hendrik Casimir to drive him to The Hague, so that he could give the Minister of War some helpful advice on defending the country. (Casimir was a theoretical physicist and the supervisor of the Kamerlingh Onnes Laboratory from 1936 to 1942, when he took a position at Philips Research.) As a member of the committee for applications of physics in warfare, de Haas was well connected in government circles, and he had the papers he needed to pass through the military checkpoints on the way to The Hague. But when the two scientists arrived, they could find no one to talk to. Casimir had no idea what de Haas was up to, and later remarked, 'It must have been very clever, but totally impracticable under the circumstances'.[11]

At Christmas 1938, Otto Hahn and Fritz Strassmann discovered nuclear fission in Berlin. Soon after, Lise Meitner and Otto Frisch publicly announced that this process released a relatively large amount of energy. Eliza Cornelis Wiersma, a professor at Delft Technical University, discussed the implications with Frédéric Joliot-Curie in Paris and, upon his return, told his former professor de Haas about the first, failed French attempts to set off a chain reaction. De Haas realised at once that this was a rare opportunity. He approached the Minister of War and was granted a meeting with Prime Minister Hendrik Colijn just a day later. As a result, ten tons of Congolese yellowcake were secretly purchased from the Belgian company Union Minière. Yellowcake is the yellow powder that remains when thorium is extracted from uranium ore. In the summer of 1939, 200 barrels of the material were delivered to the Kamerlingh Onnes Laboratory,[12] where they were stored in the cellar under the watchful eye of de Haas. He reportedly joked, 'Don't stack them on top of each other, or they'll blow', a reference to the concept of critical mass that underlies the atomic bomb.[13] Shortly before the war broke out the uranium was moved to Delft, where it was hidden in a walled-up cellar and so, according to de Haas, 'escaped' the attention of the Germans.

The Early Days of the War

After the war (and occupation) began for The Netherlands, on 10 May 1940, the top priority for the Kamerlingh Onnes Laboratory was not to upset the occupiers needlessly. The Germans did not seem to bear the lab any ill will. De Haas later claimed that this was thanks to Keesom, who soon after the invasion, 'without consulting with anyone', had written 'to ten German physics professors (including Planck) and asked them whether they thought that work at the cryogenics lab could continue under the circumstances'.[14] They replied in the affirmative and 'in a spirit of friendship'. According to de Haas, Keesom then showed these letters to the 'Sudeten German' Walter von Stokar, a professor of prehistory at the University of Cologne appointed to the education department in The Hague to investigate how Leiden University could be transformed into a German 'frontline university' organised in a military fashion.[15] In the same interrogation, held in 1944, de Haas said that Keesom had since become 'manic-depressive', and that he had seen it coming years before in Keesom's 'handwriting and walk'.

Soon after the war began, a note was posted on the bulletin board at the lab: 'The entire staff of the Kamerlingh Onnes Laboratory is emphatically urged to refrain from any action directed against the occupying power, or that could be interpreted as such by that power. … We must conduct ourselves fittingly. A civil question, no matter who asks it, should receive a civil answer'. This was followed by another note, which reiterated that 'any political propaganda, of any nature whatsoever, is strictly prohibited within the laboratory. Very serious measures will be taken if this prohibition is violated'.[16] Both of these edicts were issued by the two directors, Keesom and de Haas.

The next note was penned by deputy director August Crommelin: 'There have been incidents in which schoolchildren damaged notices posted by the German authorities or jeered at German soldiers on the street. I do not suspect any of the pupils [at the lab's school for instrument-makers] but nevertheless would call it to your attention that such things can be extremely dangerous and have extremely grave consequences, not only for the schoolchildren but for the entire citizenry'.[17] Previously, Crommelin had emphasised to the children that they were strictly forbidden to take pieces of crashed aeroplanes.

It did not take long for the persecution of Jews to begin. Jewish instructors and other staff members at the university were relieved of their duties on 23 November 1940, and Professor Rudolph Cleveringa, the dean of Leiden's law faculty, gave a protest speech at the Academiegebouw (the main university building) that would go down in Dutch history. This resulted in the closure of the university the very next day. A month earlier, Crommelin had informed Leiden's mayor and aldermen that 'none of the teachers at the Leiden Instrument-Makers School … nor their spouses or fiancées, nor any of their parents or grandparents has ever belonged to the Jewish religious community'.[18] But the 'outstanding' instrument-maker in training, Henri ('Hein') Roos, the son of a butcher from Meppel, was no longer welcome at the school as from 1 September 1942, because Jews were required to attend Jewish schools. Crommelin protested to the education department that the instrument-makers school in Leiden was unique and Hein could not continue his training anywhere else, but to no avail.[19] Hein was sent to Westerbork transit camp and died in Auschwitz on 30 September 1942.[20]

Staying Open to Prevent Theft

De Haas saw it as his mission to keep the Germans from plundering the lab. When the university closed on 28 November 1940 and the Kamerlingh Onnes Laboratory, including the institute for theoretical physics, cancelled its classes and practicals, Crommelin hastened to add that work would go on 'as usual' at the laboratory and the associated school. This extended to 'all those in the laboratory who receive pay on any basis whatsoever, and all the teachers and pupils at the school'.[21]

De Haas later came under criticism for having convinced the Germans to allow the Kamerlingh Onnes Laboratory to remain in operation. He and his co-director Keesom drew additional ire by not participating in the collective resignation of most of Leiden's professors, including the theoretical physicist Kramers, as well as Casimir and Adriaan Fokker (both of whom held endowed chairs). De Haas defended his actions to interrogators at the Military Office for Scientific Intelli-

Figure 4.1 Jaap Kistemaker in the Kamerlingh Onnes Laboratory, 1943.
Source: Leiden Institute of Physics

gence (*Militair Bureau Wetenschappelijke Inlichtingen*, or MBWI, an agency of the Dutch government in exile) with the argument that he had a weightier responsibility than other professors. If he closed the lab, he would leave behind six or seven hundred thousand guilders worth of equipment for looters.

De Haas was able to convince the Germans that the instrument-makers school had become an integral part of the laboratory. The pupils (about 100 in number during the war) were called on to participate in physics experiments and took classes (both practical and theoretical) from technicians and assistants on the laboratory staff, and the laboratory environment (with a glass-blowing workshop, electrical equipment, a drawing office, etc.) guaranteed a multifaceted education.[22]

Because the school (as a socially important institution) was allowed to remain open, the laboratory could do the same. De Haas even persuaded the German authorities to let him appoint four new research assistants.[23] And so, during the war the 'blue-collar boys', as the trainee instrument-makers were known because of the colour of their smocks, went to school as usual; this had the advantage that they were exempted from *Arbeitseinsatz*, forced labour for the Germans. De Haas also said later that he had 'stamped papers illegally' so that his 'assistants would be left alone during razzias [round-ups of Jews and other people in hiding]'.[24]

De Haas' independent course with regard to the authorities strained his relationship with Casimir, who wrote that the director's characteristic approach 'sometimes put me in a somewhat painful position at faculty meetings, where de Haas hardly ever made an appearance and I had to defend his point of view'.[25] What is more, when de Haas returned from sick leave that had begun in the autumn of 1940, he summarily rejected Casimir's proposed research plans without any discussion. Exasperated, Casimir left for Philips in April 1942. De Haas also bore a grudge against Casimir, who he believed had not kept him adequately informed while he was in the Veluwe region recovering from his illness. He accused Casimir of 'wanting to play the boss' and of taking a job as the head of the radiography department at Philips in the full knowledge that this department 'had to work almost exclusively for the Germans'.[26]

Once de Haas had played his trump card, the instrument-makers school, to keep the Kamerlingh Onnes Laboratory in operation, he needed to make sure that the closure of Leiden University would not deprive the lab – 'the world's largest and best equipped in its field' – of its students. 'Preventing this laboratory from wasting away and dying is not a Leiden issue but a national issue', he said. 'And that danger is present. When at the top its research staff can find positions elsewhere in the country, while at the bottom there is no fresh supply, a major laboratory such as the Kamerlingh Onnes Laboratory is irrevocably doomed to die. Already, a deep malaise hangs over it'.[27]

De Haas tried to solve this problem by joining forces with Delft Technical University. In September 1942, the Kamerlingh Onnes Laboratory was placed under that university's auspices. For de Haas, one reason to work with Delft was the 'incontrovertible fact that traditional university institutions are further removed from the people than technical universities'. He added, 'The placement of students from Delft in the Leiden laboratory for research purposes can best be compared to sowing a new variety of seed on a good old field. ... This is why I consider this proposal so beneficial, because when the Kamerlingh Onnes Laboratory returned to its birthplace in Leiden [after the war], two different ways of thinking, those of physics and physical engineering, encountered one another'.[28] In negotiations with Leiden, Delft was 'not very forthcoming' and de Haas came away with the impression of 'unpleasantness'. But thanks to Eliza C. Wiersma, a physics professor in Delft who had written his doctoral thesis under the supervision of de Haas and knew the laboratory in Leiden like the back of his hand, the partnership advocated by de Haas became a reality after all.

In practice, this meant that Wiersma became the head of the Kamerlingh Onnes Laboratory. But de Haas and Keesom (who remained at home on sick leave for the rest of the war) were still the co-directors, at least in title. They took care of financial matters and made proposals to the Board of Governors about budgets, appointments and so forth. They were each assigned three assistants and a technician so that they could continue their scientific work. All the other assistants were assigned to Wiersma, for research and to aid students from Delft.[29]

Again, the Kamerlingh Onnes Laboratory had escaped closure. Liquid helium was available almost every week for cryogenic research into thermodynamics (by Keesom's group) and the electrical and magnetic properties of superconducting metals (by de Haas' group). Dozens of articles were submitted to both *Physica* and the laboratory's own journal, *Communications*; articles submitted after 1943 did not appear in print until 1946. Because some research took place in the evening or at night (when there were fewer vibrations caused by passing cars and the like), de Haas obtained special passes so that the assistants and technicians involved could walk the streets after curfew (10 p.m.).[30]

But the war took its toll. Research assistants, technicians and pupils at the instrument-makers' school regularly took the afternoon off to make the rounds of farms and illegal markets for food. In the summer of 1944, gas and electricity were in short supply and the laboratory had to shut down. By this point, de Haas had fled to England, while Wiersma had been laid low by severe rheumatoid arthritis and died on 31 July after a few months' illness. In the final year of the war, Deputy Director Joost van den Handel (appointed as Crommelin's successor as from 1 January 1944) and theoretical physicist Kramers took charge of the laboratory.

A Flawed Requisition

In the meantime, the occupiers had been investigating an array of possibilities for using the 'famous' Kamerlingh Onnes Laboratory to further 'the German cause'.[31] This began when the chemist and *SS-Obersturmführer* Alfred Böttcher visited the lab in February 1943. De Haas gave him a tour of the laboratory at his request, but dodged his inquiry about coming to work there (Böttcher later found a place at the Inorganic Chemistry Laboratory).[32] In June 1943, Böttcher proposed to Abraham Esau, president of the *Physikalisch-Technische Reichsanstalt* (PTR; the German institute for science and technology) in Berlin, that the Kamerlingh Onnes

Laboratory should be placed at the disposal of German science.[33] Hans Schwarz, a *Ministerialrat* (high-level German official) based in the Dutch town of Apeldoorn, expressed similar views, claiming that the laboratory had no equal in Germany. 'The institute in Leiden', Schwarz wrote on 24 June 1943 to his fellow official Friedrich Wimmer, an SS *Oberführer*, 'must therefore be regarded as being of primary importance to the war effort'.[34]

Böttcher received instructions to take stock of the equipment and the research in progress at the Kamerlingh Onnes Laboratory. German bureaucrats were circulating plans to exempt laboratory staff from *Arbeitseinsatz*, because 'The institute will very soon become an official secret of the Reich [*eine geheime Reichssache*]'.[35] The facilities in Leiden regularly received German visitors. While bombardments were taking an increasing toll in Germany, the German occupying forces in The Netherlands continued to insist that Dutch university institutions should remain intact. In the spring of 1944, Wimmer expressly prohibited the requisitioning of items from laboratories in Leiden,[36] but gave Esau, the PTR president, permission to found a German research institute in Arnhem (this was later changed to Doetinchem) for 'data of importance to the war effort', using equipment from the Kamerlingh Onnes Laboratory. The plan was carried out in May.

On 5 May, the management of the laboratory was informed of the situation. The next day, Böttcher came by to make a list of instruments and machines to be supplied.[37] According to the requisition, the equipment was on loan and would not be transported to Germany.[38] The long wish list included two large electromagnets, a quartz spectrograph and a glass one, a high-voltage generator, pumps and all sorts of tools from the workshops, including drills and lathes. At the education department, Secretary-General Jan van Dam vigorously protested the requisition. In a letter to Wimmer, he wrote that the items on the list could fully equip 'a laboratory covering practically the entire field of physics'.[39] He also complained about the inclusion of instruments of 'historical significance' that had played a role in 'decisive experiments'. Van Dam argued that the requisition was so sweeping that when the Kamerlingh Onnes Laboratory reopened it would no longer be able to offer introductory university education to students of the natural sciences and medicine. He continued:

> The Kamerlingh Onnes Laboratory and the instrument-makers school associated with it form an utterly unique, world-renowned institute in which the Dutch people rightly take pride as a cultural achievement. It represents a cultural asset of the first water, not only in The Netherlands but throughout the German Reich. The Kamerlingh Onnes Laboratory is among the institutes where valuable scientific achievements have been made continuously in recent years. The professors affiliated with it and the rest of the staff, including the instrument-makers school, have *without exception* performed their duty with perfect loyalty, and furthermore, since September 1942 it has been officially involved in the educational programme at Delft Technical University.

After noting that the requisition was in conflict with the pledge not to disrupt work at the Kamerlingh Onnes Laboratory, van Dam concluded with the statement that he 'could not possibly justify cooperating with the implementation of the measure, either to the people of The Netherlands or to myself'.

This did not help. On 2 June, Wiersma was told that the requisition 'remained in full force'.[40] The parties involved met at Wimmer's office in Apeldoorn on 15 June. Because de Haas had just fled to England by way of Switzerland, and Wiersma and Keesom were sick, Kramers represented the interests of the Kamerlingh Onnes Laboratory. On the table was a list of research projects in progress at the lab.[41] Von Stokar, who had gone to Leiden to take a look around, was suspicious about the number of projects on the list – almost forty, and remarked that during his visit he had not seen nearly so many people.[42] He also said that the only project in which the propaganda ministry would be interested was Kistemaker's, which related to warping in photographic plates during development. This was probably ironically intended, however, given that he went on to say, 'It is inconceivable that the photographic work or any other project is *kriegsentscheidend* (crucial to the war effort, literally "war-crucial") … at most it is *kriegswichtig* ("war-important")'. This was a response to a letter from Berlin in which a Max Winkler, claiming to derive his authority from Goebbels, urged that the requisition of the Kamerlingh Onnes Laboratory's equipment be withdrawn.[43] In the report on the Apeldoorn meeting,[44] the *kriegs* ('war') in *kriegswichtig* has been

Figure 4.2 Hans Kramers (middle) and Elize Wiersma (right).
Source: Leiden Institute of Physics

crossed out. Though van Dam had described the laboratory as a 'national sanctum' at that 15 June meeting, and Kramers had on that occasion called science 'a rose on a dunghill' which 'once damaged would not unfold so easily again', and the Germans had argued fiercely behind the scenes about the need of the confiscation, the bottom line was that the requisition continued to apply in full.

Even Werner Heisenberg, Germany's pre-eminent theoretical physicist, who by that time was working in Berlin on the development of nuclear power, could do nothing to change the situation. As part of a cultural exchange initiative, Heisenberg had visited The Netherlands for a week in October 1943.[45] Kramers planned the programme, which had a pronounced scientific slant and consisted mainly of meetings with Dutch theoretical physicists. At the Kamerlingh Onnes Laboratory, Heisenberg met Casimir (who had come from Eindhoven for the occasion), Keesom and Kramers, and gave a colloquium on elementary particles. There is no evidence whatever for claims that during his visit he took a good look around so that the impending requisition of equipment would meet the needs of Germany's nuclear energy programme.[46]

In fact, Heisenberg tried to stop the requisition. Towards the end of his week in The Netherlands, during a meeting with Reichskommissar Arthur Seyss-Inquart, he argued that the Kamerlingh Onnes Laboratory should remain in operation. And on 20 June 1944, when the requisition was close at hand, he told one of the individuals involved that the Kamerlingh Onnes Laboratory was 'the world's leading institute for cryogenic research', that the special equipment there 'simply could not be relocated', and that 'reconstruction elsewhere would take years'. Germany may have had shortages of just about everything, but even so it was 'undesirable', Heisenberg declared, to requisition standard items and thereby leave the Leiden institute paralysed. 'That would create a tremendous stir in the scientific community', he added, 'and work out badly for Germany in propaganda terms'.[47]

Likewise, despite a meeting between Kramers and Böttcher on 20 July, at which they touched on the point that the requisition should not entirely disrupt any of the research projects on the list, but at most lead to 'tolerable delays' in a few cases,[48] the electromagnets and other essential items were taken. The issue had by then been under discussion so long, however, that the Leiden laboratory had plenty of time to hide all sorts of things from the Germans. Technicians and assistants were given smaller things to take home, and the larger pieces of equipment (such as the X-ray machine, the high-voltage generator and the quartz spectrograph) were lugged to places as diverse as Gravensteen (a medieval monument), the Museum of Ethnology, a garage, a candle factory and a bicycle shed.[49] Another, smaller-scale requisition followed on 5 December 1944; the additional items taken included a lathe. Kistemaker later claimed that Böttcher had turned a blind eye to their schemes: 'Every one of us understood at the time that he was trying to help, to the extent that he could. You can't pull the wool over the eyes of a capable physicist like him in his own field'.[50] After the war, the new supervisor of the Inorganic Chemistry Laboratory secured the return of the most important pieces of stolen equipment,[51] which had been taken to Germany in the meantime.[52]

The Kamerlingh Onnes Laboratory was also involved in the resistance, especially in the final years of the war. The lab, which employed a staff of almost

150 researchers, technicians and trainee instrument makers, was the perfect place for inconspicuous gatherings. All the extensions and additions to the main building over the years had created a labyrinth where it was easy to hide. In that labyrinth Rob Kollewijn,[53] a student from Delft Technical University, was working (in consultation with de Haas and Wiersma) on a method of forging identity papers that were indistinguishable from the real thing, even under ultraviolet light. In addition, Kollewijn was experimenting with the production of explosives. Others were testing devices that could quickly cut through prison bars by chemical means, or making embossed stamps. There was a radio and a printing press. Wiersma was involved in several resistance groups that used the laboratory as a base of operations. There were also Jews living in hiding there; after the laboratory closed in September 1944, they were joined by about ten young staff members and students who were trying to avoid *Arbeitseinsatz*. De Haas' office became their living room.[54]

Cellastic

After the war, de Haas found himself in hot water because of his work for an organisation known as Cellastic in 1943–1944. This trading company, established in 1937, scrutinised patents in search of useful ideas for the German war industry. It had an office in Paris for that purpose with a few dozen employees, who were essentially engaged in industrial espionage. In order not to arouse too much suspicion, Cellastic retained the services of non-Germans from scientific circles, who not only evaluated patents and translated them into German, but also visited research laboratories.[55]

In 1943, de Haas was approached by Cellastic director Hans Kleiter, a man with 'the tanned features of a sailor' who was (or claimed to be) a South African with Dutch nationality, and worked as a spy for the Germans. Kleiter had a letter of introduction from Alfred Flesche, a German banker-cum-spy, whom de Haas knew personally – the two men were both on the supervisory board of the Amsterdam company, Carp, which produced refrigeration technology. Flesche had served as a go-between when de Haas' son in Amsterdam was arrested for aiding Jews. Jan de Boer, a technician at Carp, warned de Haas about Cellastic as soon as he heard about Kleiter's advances: 'Don't start down that road, my friend; that Kleiter just may be as much of a scoundrel in France as Flesche has been in The Netherlands'. The banker and yachtsman Johan R. Carp is said also to have warned de Haas about Kleiter ('a German spy and a villain').[56]

Even so, de Haas decided to do business with Cellastic. The excellent pay – 300 guilders a month, plus reimbursement of travel and accommodation expenses for his wife – certainly played a role, and the proposal must have appealed to his sense of adventure. De Haas went to Paris with his wife three times:[57] in October 1943 and in January and April 1944. In late May, after consulting with the Dutch government in exile, they fled to Switzerland and travelled on to London. There, de Haas was questioned by the MBWI, in the presence of the physicist Samuel Goudsmit (originally from Leiden). While gathering information on the German

Figure 4.3 Wander Johannes de Haas, 1878–1960.
Source: Leiden Institute of Physics

nuclear energy programme close behind Allied lines as part of Operation Alsos, Goudsmit had stumbled upon an abandoned Cellastic office in Paris.[58] It was only after Goudsmit raised the subject during the interrogation in London that de Haas began to talk about Cellastic, without mentioning the names of other Dutch nationals involved.

In Paris, de Haas had tried to sell Cellastic an invention of his for drying grass. In his role as the director of the renowned Kamerlingh Onnes Laboratory, he could easily make appointments to visit laboratories. He also visited the home of Joliot-Curie, who later told the MBWI that he had known de Haas before the war and never seen him as a collaborator. He added that during his visit de Haas had 'taken no pains to conceal his anti-German and anti-Nazi sentiments'.

One of de Haas' assignments at Cellastic was to recruit a young Dutch physicist. After a couple of failed attempts, he reeled in Jaap Kistemaker, who had become Keesom's assistant in 1942. Kistemaker could not resist the fee of 5,000 francs (his wife received 125 guilders a month) and the free trips to Paris. In February and March 1944, Kistemaker spent a total of three weeks at the Cellastic office, translating patents for automobile headlights, electron multipliers and high-pressure mercury lamps. He also took trips to industrial laboratories. When Kistemaker returned from his first trip to Paris, he reported to de Haas that suspect business was going on at Cellastic. Kistemaker also informed Wiersma, who said he would pass on the information to London through his contacts in the resistance.

While Kistemaker saw his actions as a textbook case of counterespionage, the MBWI took a different view of the situation. 'Kistemaker's observations about his German superiors and associates, all the members of the German Wehrmacht and his French associates, most of whom were collaborators, were of very little use. He says he had a very interesting time [in Paris]. Apparently, he was taken in by the generosity of men who, more than once, "accidentally" paid him too much salary'.[59] True, Kistemaker had reported his suspicions to de Haas, 'but the impression remains that, for the sake of a very pleasant life in Paris, he did not take the attitude that one may expect of a good Dutchman, in view of his background and intelligence'.[60] Goudsmit arrived at a similar conclusion, a fact which deeply aggravated Kistemaker, who later made his name as the man who broke through the American monopoly on uranium production in 1952 and helped to develop the ultracentrifuge.[61]

Conclusion

What are we to think of the conduct of Wander Johannes de Haas during the war? He seems to have been correct in his belief that the best way for the Kamerlingh Onnes Laboratory to prevent the theft of instruments and other equipment was by staying open. The occupiers hardly interfered with the laboratory's work. Not until the end of the war did they take some equipment, and even these requisitions were limited in extent. Until then, fundamental research continued as usual, and liquid helium was available well into 1944.

De Haas put on a show of good will towards the occupiers so that his strategy would succeed. He took this so far that in May 1944, when the Germans finally decided to requisition some items, Secretary-General van Dam at the Education Department protested, reminding them of the loyalty that de Haas and his staff had always shown – a dubious compliment. But no one who worked side by side with de Haas at the Kamerlingh Onnes Laboratory has ever accused him of pro-German sympathies. In short, his tactics seem to have worked well. This does not change the fact that he could be remarkably tactless towards close associates such as Keesom and Casimir.

De Haas' relationship with Cellastic deserves much sharper criticism, and financial considerations certainly played a role in it. The same can be said in the case of Kistemaker, though there is no evidence at all that he took part in uranium research for the Germans during the war, as the journalist Wim Klinkenberg has claimed.[62] The MBWI judged the two of them harshly and was surprised that de Haas was suspended so briefly. De Haas, in contrast, felt mistreated and was enraged by the 'unjust treatment' he received. His only goal, he said, had been to gather intelligence, and he had let the government in exile know that he was eager to come to London and report on his findings for the benefit of the Allied war industry.

De Haas was sometimes a bit out of touch with reality, constructing all kinds of conspiracies, a fact which accounts in part for his bitter, aggrieved attitude after the war. When his suspension as professor was lifted by the Purgation Board in November 1945, he indignantly remarked, 'It is still a great scandal. The entire process is shameful'. And in response to the generous donation of a million guilders to the Kamerlingh Onnes Laboratory on 1 March 1946 by the oil company BPM[63] (Bataafsche Petroleum Maatschappij), his characteristic remark was 'Not bad, eh, for someone who worked with spies'.[64]

Notes

1 The thesis gives W. H. Keesom as the supervisor, but Keesom was ill on the day of the defence and was replaced by W. J. de Haas. Later, however, de Haas told an interrogator at the Military Office for Scientific Intelligence (*Militair Bureau Wetenschappelijke Inlichtingen*; MBWI) that he had refused to officially award the degree because Saris was the 'leader of the NSB (Dutch National Socialist) students in The Netherlands'.

2 D. van Delft, *Freezing Physics: Heike Kamerlingh Onnes and the Quest for Cold* (Amsterdam: Edita, 2007).

3 See *Physica* 5 (1938) and 7 (1940). The articles also appeared in *Communications*, nos. 252d, 257d and 260a.

4 It was published in the November issue.

5 Many students from Leiden continued their studies in Amsterdam, Utrecht or Delft.

6 Saris' last measurements are dated 4 July 1940. See B. F. Saris, *De warmtegeleiding in vloeibaar Helium II* (Amsterdam, Noord-Hollandsche, 1941), 36.

7 P. J. Idenburg, *De Leidse Universiteit 1928–1946* (The Hague: UniversitairePers Leiden, 1978), 162–5.

8 J. Kistemaker, *Thermodynamische eigenschappen van helium in de buurt van het λ-punt* (Leiden: Ydo, 1945).

9 D. van Delft, *Heike Kamerlingh Onnes: de man van het absolute nulpunt* (Amsterdam: Edita, 2005), 523–36.

10 C. J. Gorter, *Jaarboek Koninklijke Nederlandse Akademie van Wetenschappen 1959–1960*, 300–3; A. Maas, 'Mesmerized by Onnes', In: E. Wyka, M. Kluza and A.K. Zawada (eds), *East and West. The Common European Heritage* [Proceeding of the XXV Scientific Instrument Commission] (Krakow: Jagiellonian University Museum, 2006), 225–30. See also the article by J. van den Handel in *Biografisch Woordenboek van Nederland*.

11 H. B. G. Casimir, *Haphazard Reality. Half a Century of Science* (New York: Harper and Row, 1983), 173.
12 For more information about this episode, see J. van Splunter, *Kernsplijting en diplomatie: de Nederlandse politiek ten aanzien van de vreedzame toepassing van kernenergie 1939–1957* (Amsterdam: Het Spinhuis, 1993), 26–39.
13 Casimir 1983, 203–4. Of course, this was natural uranium (in uranium oxide), with a small proportion of fissionable uranium-235.
14 De Haas made this statement to the MBWI (see note 1) on 16 March 1945. NIOD, archive 251b, no. 412.
15 Idenburg 1978, 219–22.
16 Notes from the archives of the Huygens Laboratory in Leiden.
17 Ibid.
18 Ibid.
19 Crommelin to the Ministry of Education, Science and Culture (*Departement van Opvoeding, Wetenschap en Kultuurbescherming*), 16 January 1942. Huygens Laboratory archives.
20 http://www.joodsmonumentmeppel.nl/families/roos-s/roos-s.html (last accessed 31 May 2007).
21 Statement issued by C. A. Crommelin, 28 November 1940. Huygens Laboratory archives.
22 Van Delft 2005, 298–313.
23 Personal communication (letter dated 23 April 2005), Dr H. P. R. Frederikse, one of the four assistants.
24 NIOD, archive 251b, no. 403.
25 Casimir 1983, 237–8.
26 NIOD, archive 251b, no. 387.
27 Huygens Laboratory archives, correspondence of De Haas.
28 Ibid.
29 Decision OWK no. 241, 29 August 1942. Huygens Laboratory archives.
30 Application dated 11 February 1942, Huygens Laboratory archives.
31 K. Berkhuysen, 'Het Duitse laboratorium in Doetinchem', *Jaarboek Achterhoek en Liemers 2006*, 36–56.
32 NIOD, archive 251b, no. 412.
33 NIOD, archive 20, file 558, Böttcher to Esau, 20 June 1943. In 1939 Esau was one of the initiators of the German nuclear research programme, the *Uranverein.*
34 NIOD, archive 20, file 558, Schwarz to Wimmer, 24 June 1943.
35 NIOD, archive 20, file 558, Von Stokar to Ammelung, 15 July 1943.
36 Idenburg 1978, 291.
37 The Inorganic Chemistry Laboratory in Leiden and the Physics Laboratory of the Vrije Universiteit in Amsterdam also had to surrender equipment.
38 Idenburg 1978, 291–2.
39 Van Dam to Wimmer, 13 May 1944. Quoted in Idenburg 1978, 292–3.
40 Idenburg 1978, 293.
41 This list is in the Huygens Laboratory archive, Wiersma file.
42 NIOD, archive 20, file 558, von Stokar to Wimmer, 7 June 1944.
43 NIOD, archive 20, file 558, Winkler to Fischböck, 17 May 1944.
44 NIOD, archive 20, file 558, report on meeting in Apeldoorn on 15 June.
45 D. C. Cassidy, *Uncertainty: The Life and Science of Werner Heisenberg* (New York: Freeman, 1992), 470–3.
46 Idenburg 1978, 291.
47 NIOD, archive 20, file 558, Heisenberg to Fischböck, 20 June 1944.
48 NIOD, archive 20, file 558, report on Kramers' meeting with Böttcher and Grünwald, 10 July 1944.

49 Idenburg 1978, 412.
50 Afterword by J. Kistemaker in Michel Bar-Zohar, *De jacht op de Duitse geleerden 1944–1960* (Utrecht/Antwerp: A.W. Bruna & Zoon, 1967), 271.
51 Huygens Laboratory archive, Wiersma file, memorandum by J. van den Handel.
52 Shortly before the German laboratory, which was housed in a teacher training school in Doetinchem, was destroyed by Allied aircraft tipped off by the resistance, the equipment was moved to a location in Duitsland.
53 In November 1944, Kollewijn was arrested during a raid in Delft and died from the injuries he sustained.
54 Idenburg 1978, 416. Those in hiding included Dick ter Haar, Jan Berend Westerdijk, Lothar and Marion Mayer and Hans Frederikse (personal communication, Hans Frederikse).
55 NIOD, Cellastic archive, inv. no. 251b.
56 Ibid.
57 In addition to De Haas and Kistemaker, J. Ketelaar (a chemistry professor from the University of Amsterdam) and J. W. Zwartsenberg worked for Cellastic.
58 S. A. Goudsmit, *Alsos* (New York: Sigma, 1947), 43–5.
59 NIOD, archive 251b, no. 233.
60 NIOD, archive 251b, no. 234.
61 M. Traa, ‘Cellastic, reconstructie van een affaire’, *Wetenschap, Cultuur en Samenleving* (March 1996), 31–40.
62 W. Klinkenberg, *De ultracentrifuge 1937–1970* (Amsterdam: Van Gennep, 1971).
63 A Shell daughter.
64 NIOD, archive 251b, no. 349.

5 Electron microscopy in Second World War Delft

Marian Fournier

In 1949, the Philips factory in Eindhoven (The Netherlands) distributed the first examples of its famous transmission electron microscope, the EM100. Three years before, early in 1946, just eight months after the end of World War II, the board of directors of Philips had decided to build a small number of electron microscopes to test the market and to gain experience with their construction. The production of the first specimen of the lot was hurried along so that it could be demonstrated at a conference on electron microscopy in Oxford in September of the same year. The instrument was ready in the nick of time and was flown over the day before the conference began. Even though at the supreme moment it would not work (!), those attending the conference were full of praise; one verdict was that it was 'a dashing new departure in electron microscope design'.[1]

Philips had indeed put much thought into the design of their electron microscope.[2] It was made to resemble a writing desk with smooth aerodynamic lines (Figure 5.1) and, because of the tilted position of the 'optical' tube, the fluorescent screen was situated right in front of the observer, facilitating the study of details of the image on the screen. But the real innovation was hidden in the interior of the instrument. The optical tube contained no fewer than five lenses instead of the usual three and this combination allowed continuous magnification variation between 1,000x and 100,000x and selected area diffraction. In other words, with the EM100 a scientist could easily switch from a low magnification to a high magnification, with any number of steps in between, and could both study the image of a selected spot and see the diffraction pattern of that same area by simply turning a dial. How could it be that Philips were able to present such a piece of cutting-edge high technology just one year after the war?

The development of the Philips EM100 began just before the outbreak of World War II and took place mainly at the Laboratory of Applied Physics of the Technische Hoogeschool (Technical University [TH]) in Delft. The central figure throughout this period was Jan Bart le Poole (1917–1993, Figure 5.2) who, in the final years of his studies at the TH, started the construction of an electron microscope as his graduation project. The actual engineering of that first attempt lasted from September 1939 until April 1941. Only months later, le Poole started on the construction of a second electron microscope. This instrument, later known as 'Mark II', contained the essential features of the optical column of the Philips

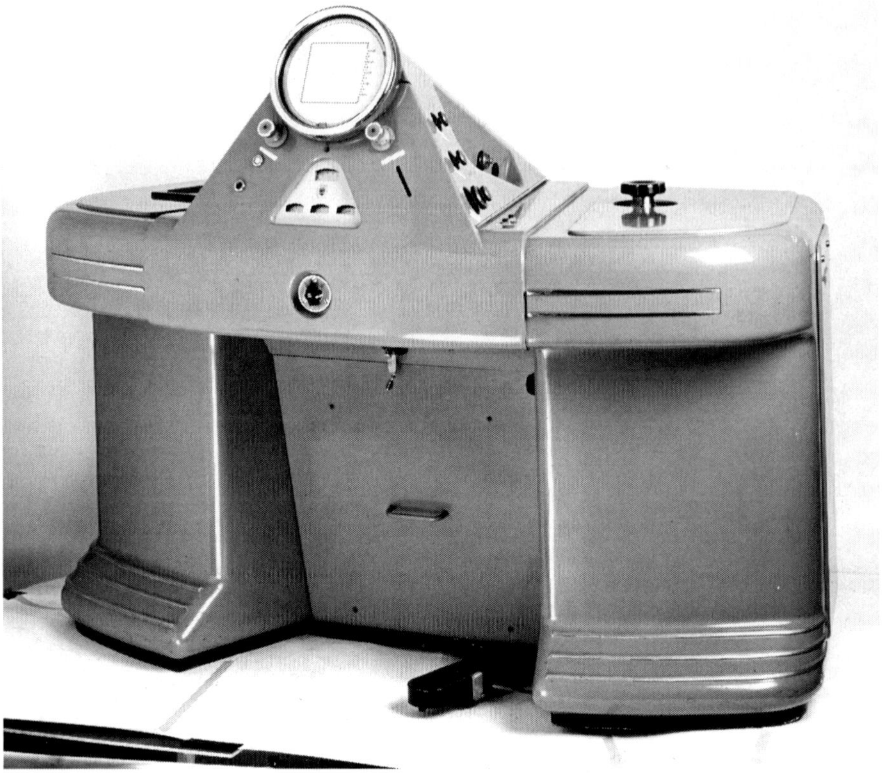

Figure 5.1 A copy of the Philips EM100, made in the 1950s, courtesy Museum Boerhaave, Leiden, The Netherlands.

EM100 that were subsequently widely considered as significant contributions to the development of electron microscopes.

The story of the events during the ten years between le Poole's proposal for his graduation project and the launch of the Philips EM100 in 1949 has been told in ample detail by some of the participants[3] as well as by fellow electron microscopists and colleagues.[4] These tales were mostly told on the occasion of some celebration or other and were primarily aimed at the community of physicists and engineers working on electron microscopes. They contain a wealth of detail concerning the engineering part of the Delft experimental electron microscopes, but other aspects, such as financing, organisation, learning how to use the instrument, collaboration with scientists from other disciplines and the uncertain circumstances imposed by the ongoing war, have received little attention.

This paper will investigate the history of the Delft electron microscope from three major viewpoints: the engineering achievement, the financial support received from Dutch industry and the influence of wartime conditions on the course of events. Apart from the restrictions, dangers and shortages that affected

Figure 5.2 Jan le Poole in the 1950s, with an adapted version of the EM100, courtesy Museum Boerhaave, Leiden, The Netherlands.

all citizens in The Netherlands, scientists also suffered limited access to scientific literature from abroad, little or no contact with colleagues involved in similar projects elsewhere in the world, as well as shortages of specific materials. It is well known that the Dutch universities and the industrial companies, in fact Dutch society in general, initially attempted to continue business and life much as normal during the German occupation: people adapted to the new circumstances.[5] The fact that many industrial enterprises benefited greatly in the early years of the war from the German intention to enlist the Dutch economy for their war efforts no doubt facilitated that adaptation.[6] However, within three years the situation gradually changed into one of enmity between the Dutch populace and the German authorities. That change was due mainly to the heavy demands of the Germans as regards the production of goods and labour for the German war machine, which resulted in a serious reduction of the conditions of daily life. Moreover, by that time it also became clear that the Allied Forces were going to win the war. And so, in an atmosphere of slowly mounting tension and turmoil the electron microscope was developed in Delft.

Overture: Mark I

In the summer of 1939, Jan le Poole, who had started studying in Delft in 1936, suggested to his supervisor, H. B. Dorgelo,[7] that he construct an electron microscope as his graduation research project. Many years later he recalled the pure enthusiasm that suggested the subject to him:

> I knew that an electron microscope existed. I dipped into electron optics, kept up with the literature a little and I found that fascinating, so I thought 'I'd like to do that too.' Yes, a nice subject for graduation.[8]

From 7 April 1931, electron microscopes had fairly soon attracted interest worldwide. It was on this day that Ernst Ruska and Max Knoll (at the Technische Hochschule) in Berlin made the first micrograph with the first ever electron microscope.[9] It was a rather primitive instrument: they only acquired a magnification of some 17x, but in December 1933 Ruska's subsequent work resulted in an instrument that produced magnifications of up to 12,000x. From 1937, Ruska and Bodo von Borries were engaged by Siemens & Halske to develop commercial electron microscopes and in 1939 the first of these was in operation. By the late 1930s the work of Ruska and von Borries was being emulated worldwide by physicists working in university or industrial laboratories. Even so, most of the scientific literature concerning electron optics and the electron microscope stemmed from German laboratories. Apart from Ruska and von Borries working on an electron microscope with electromagnetic lenses at the Siemens laboratory in Berlin, another group, headed by Ernst Brüche, was employed by the Allgemeine Elektrizitäts Gesellschaft (AEG, also in Berlin) developing an electron microscope with electrostatic lenses; Manfred von Ardenne's research on electron optics provided work for yet another group.[10]

In The Netherlands several laboratories kept tabs on the development of electron microscopes, including the Laboratory for Applied Physics in Delft and the Philips Natuurkundig Laboratorium (Physics Laboratory, Natlab for short) in Eindhoven. At the TH in Delft electron optics was one of the subjects that the students reading applied (or technical) physics studied and explored. At the Natlab an emission electron microscope for metallurgical work was built and used by W. G. Burgers and J. J. A. Ploos van Amstel between 1935 and 1940.[11] The development of the electron microscope also attracted the attention of Dutch daily newspapers and popular science journals.[12]

Although this paper deals almost exclusively with the two electron microscopes built by le Poole during the war, Mark I and Mark II, early electron microscopy in The Netherlands was not limited to these microscopes. At the same time a 400 kV high-voltage electron microscope was being built at the Philips Natlab in Eindhoven and another transmission electron microscope, but in this case with electrostatic lenses, was under construction in Delft.[13]

At the time le Poole suggested his subject for graduation to Dorgelo he did not know, or so he has always claimed,[14] that the latter had only recently returned from a visit to Berlin in the company of F. G. Waller Jr,[15] deputy director of the Nederlandse Gist- en Spiritusfabriek (Dutch Yeast and Spirits Factory [NG&SF]), and Professor A. J. Kluyver,[16] head of the Laboratory for Microbiology at the TH. The director of the Nederlandse Siemens Maatschappij B.V., Jonkheer Ir. A. S. C. Stoop van Strijen, had arranged for Dorgelo and his companions to visit the Berlin laboratories of Siemens & Halske to examine and sample their newly produced electron microscope.[17] Such visits were part of the intense competition between Siemens & Halske and AEG concerning their electron microscopes.[18]

As a result of their visit Dorgelo, Waller and Kluyver actively worked together throughout the 1940s and 1950s to arrange support, financial and otherwise, for the development of the electron microscope. The three men were spurred by different interests. Dorgelo had in 1928 been appointed as the first professor in applied physics at the TH in Delft. He had come to Delft from the Philips Natlab and was an ardent advocate of practical research at the university commissioned by industrial parties. To this end, he initiated the establishment of the Technisch Physische Dienst TNO-TH Delft (Technical Physics Service [TPD]), which was formally established in 1942.[19] The TPD was a partnership between the TH and the Organisation for Applied Scientific Research (TNO). Throughout World War II the TPD was housed in Dorgelo's Laboratory for Applied Physics and as a result of the grim conditions during the war remained a fledgling organisation. As Dorgelo had worked on X-rays during his time at the Natlab, he had kept abreast of Ruska's work and one of the subjects on his curriculum was electron optics. Developing the electron microscope consequently combined two of his foremost interests and also became one of the major concerns of the TPD.[20]

The interests of Kluyver and Waller, on the other hand, were in the field of microbiology. From the mid-1920s, microbiology professor Kluyver, who was engaged as consultant to the yeast factory NG&SF, and Waller conferred regularly about the scientific problems raised by the organic production processes on which the welfare of the yeast factory depended.[21] Both Kluyver and Waller naturally set great store by up-to-date and detailed knowledge of the microorganisms used in the production process. As for yeast, an important question at the time was whether these organisms had a nucleus, like the cells of higher organisms. The nucleus could not be demonstrated with the light microscope, but the electron microscope seemed to offer a new means of doing so.

The three men were much impressed with the results of the demonstration in Berlin, although Waller was slightly disappointed because the yeast cell had appeared as a uniform black blot in the electron microscope. Buying such a microscope was out of the question as the price, 70,000 to 80,000 Dutch guilders (equivalent to some €630,000 in 2007), was beyond their means. Therefore, in the report describing their experiences with the Siemens electron microscope, they advocated a plan to bring together a group of interested parties, buy one microscope jointly and share time for research with the instrument.[22] They reckoned that other Dutch companies such as De Staatsmijnen (DSM), Heineken's Bierbrouwerijen and the Bataafsche Petroleum Maatschappij (BPM, a subsidiary of Shell) might be interested in the proposition as well. Reputedly, the three men also toyed with the idea of one of Dorgelo's students possibly building an electron microscope.[23] It is no wonder, therefore, that when Jan le Poole proposed his plan for an electron microscope to Dorgelo the latter agreed immediately.

At the time, the direct exchange of information among scholars, at conferences for example, had not yet been seriously affected by the looming war, at least in Western Europe. So, in February 1940, a small meeting on electron microscopy, probably a spin-off from the Berlin visit and encouraged by the current work on lenses, was organised in Delft by the Stichting Biophysica (Biophysics Foundation).[24] Helmut Ruska, Ernst Ruska's brother, was the guest of honour and he was to lecture on his work with the electron microscope. Helmut had from the outset been convinced that the electron microscope would become an important new tool in the investigation of biological objects. In the Siemens laboratory Helmut worked with several collaborators to prove his point.[25] From 1938 to 1940, he was the first to publish micrographs of virus and phage particles made with the electron microscope, results that went largely unnoticed owing to the isolation of Germany at the time.[26] Helmut, being trained as a physician, probably could not tell his audience in Delft much about the technical details of the Siemens electron microscope, but he did present an illustrated overview of the investigations being done with the electron microscope in the Siemens laboratory and, more importantly, of the techniques used in preparing specimens. At the conference, Dorgelo and his colleague, R. Kronig (who taught theoretical physics), presented papers on various sorts of electron microscopes and the principles of electron optics respectively. Professor W. G. Burgers demonstrated the metallurgical emission electron microscope that had been built by Philips in around 1935 and that he and Ploos van Amstel had been using for some years. Helmut's lecture surely strengthened the expectations in Delft as regards the prospects for their own electron microscope.

Soon after le Poole started working on his microscope, in the autumn of 1939, he was offered a post as assistant to Dorgelo, which was a salaried post. This position, which he kept until September 1941, proved to be providential. In December 1940, by which time The Netherlands had been living under Nazi occupation for six months, the TH was closed by the German authorities until April 1941. The shutdown was the direct result of the protest by the students in Delft, as at other Dutch universities, against the dismissal of Jewish professors.[27] In Delft, all teaching was suspended for four months, but the employees of the university were allowed to continue their work. So le Poole, although still a student, but being an assistant to Dorgelo, could continue work on his graduation project.

Le Poole was busy conducting experiments and designing and constructing his first electron microscope for about 18 months. The first micrograph with the finished instrument, which was afterwards referred to as 'Mark I', was taken on 16 April 1941 and came up to expectation.[28] It was a 40 kV, two-stage electron microscope, fitted with electromagnetic lenses, magnifying some 10,000x. The construction of the instrument closely resembled the scheme for Ruska's 1933 electron microscope and the first Siemens electron microscope, built in 1938. In other words, it was a conventional scheme, based entirely on facts available from published sources.

Inspiration: Mark II

The promise of Mark I was such that le Poole, after graduating in the early summer of 1941, was engaged by Waller for the NG&SF to improve it.[29] But that was not his only task. Many years later he recollected:

> After graduation I found myself at the Yeast Factory to upgrade that electron microscope. But the interest in the project was somewhat limited as at the same time I had to solve another problem that troubled the Yeast Factory. That was capturing the smoke and soot from the gas generators that replaced gasoline during the war. The gas generator was a kind of stove that was towed behind a car and which produced gas for the car.[30]

So Waller had two purposes in mind when he employing le Poole. One was to solve an immediate and pressing problem for the NG&SF: protecting their yeast from dirt during transport to bakeries all over the country. And the other was looking to the future: the development of the electron microscope, which might prove to be an important new scientific instrument. For the time being, le Poole was to do his work in the Laboratory of Applied Physics.

Le Poole's new employer was, as far as the electron microscope was concerned, mainly interested in being able to study the details of yeast. However, the electron beam produced by the 40 kV electron microscope was unable to penetrate relatively thick biological specimens such as yeast cells; they merely appeared as black blots in the micrographs. Therefore, the object of his work had to be the construction of a microscope with a stronger electron beam, which meant that in the improved version of Mark I the voltage had to be raised. In order to acquire a high-voltage energy source, and particularly one with a stable output, le Poole made contact with A. C. van Dorsten,[31] then working at the Philips Natlab and responsible for high-voltage research and development. Van Dorsten had been an assistant to Dorgelo before he began working at the Natlab in 1937. From February 1942, van Dorsten and le Poole were in regular contact, a collaboration that in future times would culminate in a number of fine electron microscopes.

In early 1942, a high-voltage supply for le Poole's new microscope was ordered from Philips. The price was estimated at fl5,000 (a sum equivalent to more than two years' salary for a university-trained man like le Poole), but that proved to be a miscalculation as in the end it cost twice as much.[32] The generator was delivered in April 1943.[33] The circumstances of the war influenced the production of this apparatus to some degree. Many years later Van Dorsten recalled:

> Wartime shortages show up when we find in the records that the design was modified in order to reduce the quantity of transformer oil needed. The oil, to be supplied by the customer by terms of agreement, appeared to be of poor quality and caused a breakdown at 95 kV, however, without damaging the equipment.[34]

Also, in December 1942, the Philips factory was bombed by the Royal Air Force. The bombing caused some damage to the Natlab and therefore the projects at hand suffered some delay, but work could be resumed quite quickly.

While van Dorsten was busy developing a 150 kV electron source, le Poole began experimenting with Mark I: the voltage was raised to 70 kV with equipment borrowed from Philips.[35] As le Poole had by then also realised that employing a much higher voltage for the electron gun would mean that a conventional two-stage column would be much too long to work with comfortably, he decided that a three-stage column would be preferable. He realised this idea at short notice by simply sawing Mark I into two halves and inserting a third lens and found the result much to his liking.[36] The in-between lens was an original idea that subsequently proved to be a crucial step in the further development of electron microscopes, particularly as the third lens made continuously variable magnification possible as a by-product.

Throughout the war, the Siemens and AEG laboratories published widely on and promoted their electron microscopes.[37] As scientific literature (at least from within Europe) was available until the end of 1942, the small group in Delft kept more or less abreast of developments in Germany. In addition, von Borries visited Delft in late 1942 to lecture on recent developments concerning the Siemens electron microscopes. It was a visit with far-reaching consequences. One of the slides von Borries showed his audience made le Poole realise that selected-area diffraction could also easily be achieved with the in-between lens he had just incorporated in Mark I. In retrospect, le Poole recalled his brainwave as follows:

> He [von Borries] showed a slide with the ray paths, in which the rays leaving the object ran through the centre of the objective. Since I knew that it was only with some care that one could align the shadow image of the objective aperture around the projector bore, I was very well aware that the rays in his slide did not actually exist. Or could they perhaps sometimes? It then struck me that diffracted rays would show this sudden change of direction, and with that, selected-area diffraction was re-invented.[38]

As soon as possible, le Poole experimented with his idea and as a result another original lens was invented. This lens, the diffraction lens, was incorporated into the design for Le Poole's new electron microscope.

At the time of von Borries' visit, the three-stage version of Mark I was in operation. It was now connected to a 70 kV electron generator and worked well. But as von Borries had chosen to pin the Nazi swastika to his clothing, le Poole was disinclined to inform him about his results or to show him the machine. He therefore told von Borries that the microscope was dismantled for necessary upgrading.[39] Even so, he could not refuse to show von Borries parts of his microscope and some micrographs. Von Borries professed himself to be impressed.

So le Poole developed two important new ideas in the course of 1942: three-stage magnification and selected-area diffraction, for incorporation into the design of his second electron microscope, Mark II. He realised them by adding two lenses to the usual three lenses in a two-stage electron microscope. They were the diffraction lens and the intermediate lens and were situated between the objective and projector lenses. Continuously variable magnification (between 1,000x and 60,000x) was realised by gradually changing the excitation of the various lenses. Therefore, in Mark II, the smallest magnifications with the electron microscope are linked to the largest of the light microscope, by means of an original arrangement of four lenses between the specimen and the final image. The Delft group claimed in several publications that this link was a huge help in the interpretation of the images.[40] The diffraction lens, in combination with a selected-area diffraction aperture, produced diffraction patterns of very small sections of a specimen. By switching off the diffraction lens, the electron microscopic image of exactly the same section became visible on the image screen.

The Instituut voor Electronenmicroscopie (IEM): Funding and Organisation

By 1942, as the costly high-voltage supply for le Poole's new microscope was about to be ordered from Philips, Dorgelo, Waller and Kluyver realised that they were going to need some further financial support for the enterprise, especially as they thought that it was necessary to appoint an assistant for le Poole. The three men decided to turn to the Delft University Fund. The several drafts Kluyver and Dorgelo discussed for the application for the grant[41] are of particular interest. In the first draft Dorgelo suggested asking for a subsidy to appoint a technical assistant, since they had now progressed to the stage where they could start fundamental, especially microbiological, research. Kluyver observed that it would be better to appoint a biologist, in view of the research that was foreseen. Kluyver also discussed the issue with Waller and finally the three men agreed that they would ask for two assistants to be employed, but delay appointing them until later in the year, so that in the first year they would only need the equivalent of one year's salary for one assistant. The University Fund accepted these applications and so, in the autumn of 1942, le Poole was joined by Harry de Heer and Anton Quispel.[42] De Heer was engaged specifically to develop suitable photographic equipment and techniques and Quispel was to start biological research with the electron microscope.

Earlier, Dorgelo, Kluyver and Waller had also rekindled the plan they had discussed on the way back from Berlin to purchase and use an electron microscope together with a group of interested parties. Buying a microscope was no longer necessary now that le Poole's upgraded Mark I was in working order and a second microscope was being built. But sharing the cost of the further development of the instrument, of developing techniques for specimen preparation and of making a survey of promising research topics was certainly still desirable.

It was an attractive plan, but difficult to realise in wartime conditions. The main problem was that the most fitting legal construction, a foundation, could not be used, owing to the laws enforced by the German occupation.[43] The best construction in the eyes of their legal adviser A. C. J. Mulder, the Philips company lawyer, was a simple contract. But Waller opposed that idea as he was afraid that the industrial partners would lose their influence on the project.

In the autumn of 1942 the negotiations received extra impetus from the previously mentioned visit that von Borries paid to Delft. It is reported that von Borries:

> was enthusiastic about the way some technical problems were solved and he advised Dorgelo to patent these as soon as possible. He even declared that Siemens would be willing to buy these patents for Germany.[44]

Of course, the question of the future patents that might result from le Poole's work was an important issue in the dragging negotiations. Until 1942, only Dorgelo's laboratory and Waller, mainly Waller, had invested in the project, but from the time van Dorsten became involved in the project, Philips too began supporting it. To start with, they designed and produced a generator for the microscope and from May 1943 they provided for le Poole's wages.[45] The parties involved finally agreed that any patents resulting from le Poole's work before November 1943 would belong to the yeast factory and Philips jointly and that Philips would acquire the patents resulting from his work after that time. But in fact no patents were acquired until after the war.[46]

Finally, after two years of negotiations, the plan was realised on 1 November 1943 with the establishment of the Instituut voor Electronenmicroscopie (Institute for Electron Microscopy [IEM]) in Delft.[47] The IEM was formally part of, but financially largely independent of, the TPD.[48] It was supported by several Dutch companies: the NG&SF, of course, Philips-Van Houten (a pharmaceutical subsidiary of Philips), Algemene Kunstzijde Unie (AKU) and Heineken, and also by TNO. The industrial partners naturally had an interest in the electron microscope that was based on their commercial pursuits: the production processes of both the NG&SF (yeast, several bacterial products) and Heineken (beer) was dependent on microorganisms. And AKU produced very fine artificial fibres, one of the objects frequently studied with the electron microscope from its very beginning.

By that time, 1944, liberation from the German occupation was a glimmer on the horizon and therefore the future interests of these companies probably played an important role in the decision to partake in the IEM. The same prospect may have inspired Unilever to join in the autumn of 1944 and strengthened the board of advisers of the IEM in their expectation that more companies, especially BPM and Hoogovens, would become involved once the war had ended.[49]

Even though people were looking ahead to the end of the German occupation, the uncertainty for the future was intense. Driven by that insecurity, I imagine, Dorgelo wrote a rather perplexing letter to G. Holst, the head of the Natlab, wondering whether it would be fair to Siemens to exploit the Delft electron microscope as it was derived from knowledge originally developed by Siemens. Dorgelo also wondered whether it would be possible to operate the IEM with the consent of Siemens and with the option that Siemens would buy the patents for the improvements developed in Delft.[50]

Immediately after the war BPM, as expected, De Staatsmijnen (DSM) and Organon joined as sponsors of the IEM.[51] But only a few years later a new structure for financing research and development was established by the Dutch government and electron microscopy was one of the subjects that profited greatly from the new funding.[52] As a result, the IEM was suspended as a separate entity in 1950, but the work and the employees continued as a division of the TPD.[53]

The Upshot

A year before the IEM was formally established, in October 1942, Quispel began his research on biological objects.[54] He was stationed at the Laboratory for Microbiology, but had to go to the Laboratory of Applied Physics to examine the specimens he had prepared. In view of Waller's interest in the nucleus of the yeast cell, Quispel began to work on that. As yeast cells did not permit the passage of electrons accelerated by the 70 kV electron microscope he had at his disposal (Figure 5.3), he had to devise a means of making details visible. Two courses seemed possible: to stain the nucleus with an appropriate stain or to make the protoplasm more transparent by removing the proteins. In fact, Quispel combined both procedures, but that was not enough to make any structures visible within the cell. The yeast cell still appeared as a black blot in the micrographs.

Just six months after joining le Poole's group, Quispel's efforts were cut short by the circumstances of the war. In April 1943, the German occupation ordered all students to sign a declaration of loyalty. Those who did not risked being recruited for the *Arbeitseinsatz*. At the same time, the German authorities dictated that all graduates must work in Germany for some time. As Quispel had only recently completed his academic education, he ran the risk of being sent to Germany. Moreover, Kluyver estimated that working with the electron microscope might not appear serious enough to the Germans to safeguard a man from enforced labour. These circumstances compelled Quispel to resign from his post.[55]

Kluyver then offered the job to Woutera van Iterson, a biology student from Utrecht. In an autobiographical document[56] she explains that one reason she was offered the job was because she was a woman and did not run much risk of being sent to Germany. As van Iterson was not in very good health at the time, she asked a fellow student, Jeanne van Brakel, to share the job with her.[57] And so the two women took up the work Quispel had left behind. Apart from staining and extracting material from the cells, van Brakel tried experiments with the microtome, and even though she did not succeed in making ultrathin sections, she deemed sectioning the most promising method as regards preparing specimens for the electron microscope,[58] which in fact turned out to be the case.

At the same time, research was being done at the IEM on a number of subjects, to name a few: clay particles (for the Chemistry Laboratory of the Agricultural University in Wageningen), dye samples (for CIMO [Central Institute for Research of Substances] in Delft) and tubercle bacilli (for T. L. Oudendal, director of the Longlijdersgasthuis [hospital for lung patients]).[59] Some observations on silver bromide sols were also being made by G. H. Jonker. Jonker referred to these

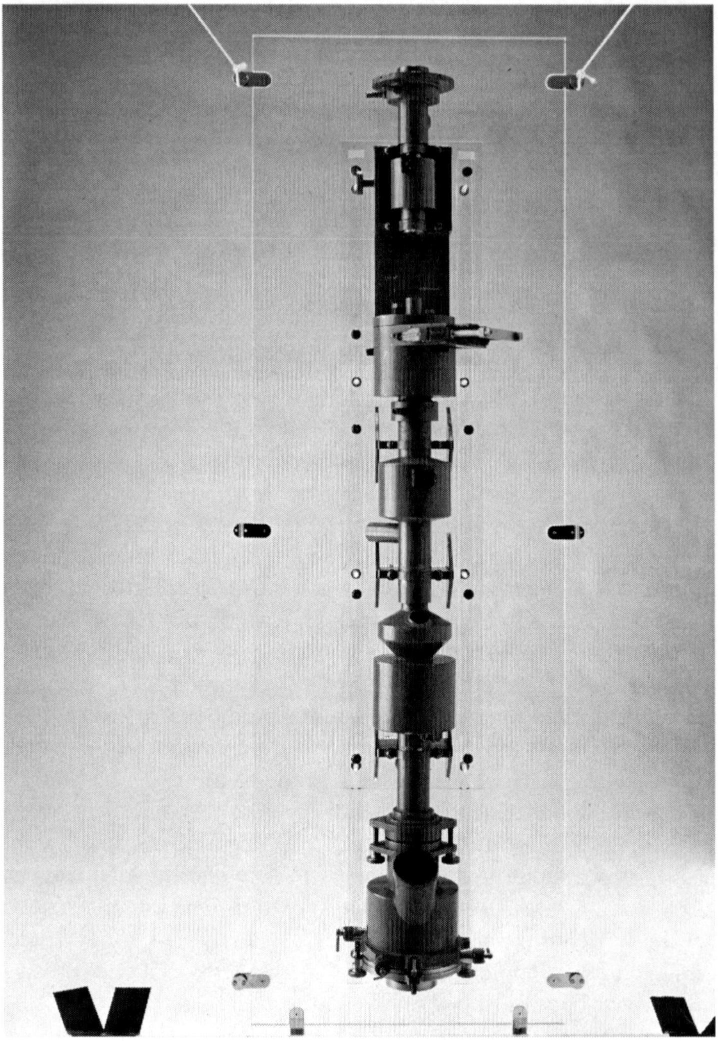

Figure 5.3 The altered version of Mark I with the in-between lens inserted in the column, in use from 1942 until 1944, courtesy Museum Boerhaave, Leiden, The Netherlands.

observations in his dissertation, which was published in 1943.[60] This is most likely the first published result with one of the electron microscopes built by le Poole.

Finally, in July 1944, Mark II was ready.[61] Le Poole recalled that:

> it worked when we turned the switch! And with our fine-grain screen we could see immediately that the image quality was much better than that of the first microscope.[62]

However, there was hardly any time to savour the resources of Mark II as by then World War II had entered a crucial stage. After the liberation of Antwerp on 4 September 1944 it seemed as if The Netherlands would be liberated within a matter of days, but in the end it was only that part of the country south of the river Rhine that was liberated. The northern part of the country faced the biggest ordeal of the war yet: a winter of famine, raids, robbery and destruction of factory equipment, means of transport and infrastructure. In Delft the TH stopped tuition altogether and it was decided to dismantle Mark II and hide its essential parts, along with many other instruments of the equipment of the TPD. They remained hidden until May 1945.

After the liberation of the northern part of The Netherlands in May 1945, le Poole and his colleagues from the IEM could start thinking about their work again. The first question to be answered was what had happened in the scientific world outside war-torn Europe? Woutera van Iterson dug up the answer. When she went to stay with her parents in the southernmost part of The Netherlands very soon after Liberation Day, she took the opportunity to read many recently published papers on the subject. Her father being the director of DSM, she had free access to the library of DSM's Centraal Laboratorium (Central Laboratory), which had recently been replenished with up-to-date scientific literature, particularly the back issues of American journals such as the *Journal of Applied Physics* and *Review of Scientific Instruments*. She prepared an extensive summary for the IEM of all the relevant papers she could find.[63] Electron microscopes had clearly been developed in the USA and in Canada during the war and had been applied to all sorts of objects. More importantly, from the literature concerning specific details of the electron microscope, it was also clear that le Poole's innovations were certainly an improvement on the electron microscopes used in the USA.[64] According to the relevant literature, variable magnification and selected-area diffraction had not been introduced by any other laboratory working on electron microscopes.[65] A similar verdict was received directly from people who had personal experience of electron microscopes. Mark II was reassembled at short notice in the summer of 1945 for a demonstration for American visitors from Shell Laboratories in Emeryville (California, USA), escorted by the director of BPM, L. H. C. Perquin. Mark II performed successfully, so much so that the American visitors professed they had not seen a versatile instrument like Mark II and Perquin immediately decided that BPM would participate in the IEM.[66]

Consequently, fortified with the knowledge that the instrument they had been working on during the war was an excellent piece of work, the group in Delft returned to their tasks and Mark II was in full working order by October 1945.[67] Over the next five years or so, the development of the Philips EM100, the development of preparation techniques for the electron microscope and fundamental and commissioned research with the microscope were to occupy a good deal of the time of the IEM. That story awaits to be told.

Conclusion

Looking back, it is clear that the development of the electron microscope in Delft proceeded without great difficulties caused specifically by the war. Not, that is, until the autumn of 1944 when, as a result of the proximity of the front line of the Allied Forces and the reprisals of the Germans for the growing resistance, public life in the north-western part of The Netherlands ground to a halt.

From the beginning, le Poole's project advanced smoothly. He was duly encouraged by his tutor, although the construction of an electron microscope went far beyond the requirements for the average graduation project. For le Poole that was no problem as he, by all accounts, was characteristically inclined to bold ventures. Moreover, his tutor, Dorgelo, firmly believed in letting students get on with their projects and see what happened. And, in hindsight, he was in this case well justified in letting le Poole do what he wanted to do.

Le Poole's first microscope held such promise that the industrialist Waller decided to support the further development of the instrument. He provided the young man with an income for almost two years and, apart from some other responsibilities, allowed him enough time to work on the electron microscope. Through the good offices of Dorgelo, Waller and Kluyver, further financial support was received from the Delft University Fund and from a number of the largest Dutch industrial companies. Le Poole's project met with approval, encouragement and financial support throughout the war years.

The conclusion that the development of the electron microscope during World War II advanced as it might have done if no war had been going on outside the front door is not surprising in view of the fact that, in general, Dutch society at first tried to continue its ordinary day-to-day business as usual and keep the war at arm's length. Both the university laboratories and the major industrial laboratories, such as the Philips Natlab, the NG&SF laboratory or Shell's laboratory in Amsterdam,[68] continued their work as best they could. Where the university laboratories experienced increasing difficulty[69] – lack of financial resources, seizure of equipment and loss of human resources – several industrial laboratories expanded during the war period. More people were employed by the Natlab in 1945 than in 1940[70] and DSM's Centraal Laboratorium, which was only established in 1940, employed, as planned before the beginning of the war, about 200 people by the end of the war.[71] From various accounts[72] relating the vicissitudes of industrial companies and laboratories during World War II it emerges that employing a surplus number of skilled men and engaging in research was a means of protecting valued personnel and preparing for the period after the war. It may be imagined that it was the same forward-looking attitude that prompted several industrial companies to support the development of the electron microscope.

During the war years the researchers in laboratories continued to work on the subjects they were already working on, but also tackled new subjects. Some of these were ordered by the German authorities, others were posed by the circumstances of the war years, as for instance le Poole's research on measures to protect yeast from soot during transport by car. But they also managed to initiate research into promising new subjects that were kept secret from the Germans. A case in point is the research on penicillin carried out by the NG&SF.[73]

Since Dorgelo had many contacts with industrial enterprises, and especially with Philips, the Laboratory of Applied Physics was in fact the ideal place for le Poole and his small group of co-workers to work on the electron microscope. Through Dorgelo's contacts, support for the project could be arranged from several industrial parties, which individually had a certain interest in the electron microscope as an instrument for research in their own laboratories, but which also had a vested interest in – and some means of – supporting research and individual scholars in dire times.

Of course, the effect of the war is clearly present in several setbacks the project encountered during the war years, but as has been indicated, they were fairly easily overcome. The most distressing incident was probably Quispel's resignation from his job with the IEM due to the declaration of loyalty demanded by the German authorities. Quispel did not return to electron microscopy after the war. His successor, van Iterson, on the other hand, made her career in electron microscopy, a subject that she might not have taken up if she had not been offered the job during the war.

Being cut off from current scientific literature and contact with colleagues elsewhere in the world was not insurmountable since the vital research, the work of Ruska and von Borries, had already been published before the beginning of the war and was available to le Poole. Moreover, as The Netherlands was occupied by the Germans, they could easily come to Leiden, as did von Borries, and so keep personal contact alive. As we have seen, von Borries unwittingly provided le Poole with a luminous idea, which to a large extent formed the basis of the success of his new design. Besides, so le Poole reflected many years later,[74] perhaps their isolation in those years might well have been a blessing in disguise. Precisely because of that isolation he had been free to pursue his own ideas, without being hampered by current notions from peers and colleagues working in the same field. Accordingly, he managed to develop, assisted by just a small number of co-workers – mainly de Heer, van Dorsten and an instrument maker employed by the IEM – a number of important innovations for the electron microscope in far from ideal circumstances. This stands in sharp contrast to the development of electron microscopes by Siemens & Halske in Germany and by the Radio Corporation of America (RCA) in the United States. In both cases, the management of the company supported the development of the microscope with adequate funds, personnel and resources within their companies and they established laboratories to explore the application of the newly developed apparatus.[75] Moreover, immediately after the war the same small group, with the support of technical staff from the Natlab, developed the first example of the future Philips EM100 on the basis of le Poole's experimental electron microscope in a matter of eight months.

There remains the question of why the German authorities did not pay any attention to the activities and achievements of the IEM. In Germany, the electron microscope was emphatically put forward by Siemens & Halske and AEG as advantageous for research related to the war effort.[76] In the USA the very same argument was used by scholars and enterprises vying for subsidies to support their work with the electron microscope.[77] Of course, no such claims were advanced by the IEM; the research in Delft was directed at ordinary subjects such as the yeast cell, tubercle bacilli, clay particles and sols. At the same time the outer appearance of both Mark I and II was definitely unassuming, if not slapdash, whereas the forty or so electron microscopes produced by Siemens & Halske during the war period were shiny, stylish machines. That may have been why Kluyver thought that the Germans would regard working with their electron microscope merely as *Spielerei*, not serious enough to protect a man from the *Arbeitseinsatz*.

Notes

1 A. C. van Dorsten, 'The beginning of electron microscopy at Philips: some historic notes', *Ultramicroscopy* 3 (1978), 118.

2 A. C. van Dorsten, H. Nieuwdorp and A. Verhoef, 'The Philips 100kV electron microscope', *Philips Technical Review* 12 (1950), 33–51.

3 J. B. le Poole, 'Über die Entwicklung der Elektronenmikroskopie in Holland', *Optik* 24 (1966), 375–82; J. B. le Poole, 'Early electron microscopy in The Netherlands', *Advances in Electronics and Electron Physics* Suppl. 16 (1985), 387–416; J. B. le Poole, 'The early days of electron microscopy in Holland', *Ultramicroscopy* 3 (1978), 113–15; Van Dorsten 1978.

4 W. van Iterson, 'Electron Microscopy in the Netherlands', T. Mulvey (ed.), The Growth of Electron Microscopy, *Advances in Imaging and Electron Physics* 96 (1996), 271–85; T. Mulvey and D. J. J. van de Laak-Tijssen, 'Jan Bart le Poole (1917–1993) Pioneer of the Electron Microscope and Particle Optics', *Advances in Imaging and Electron Physics* 115 (2001), 287–354.

5 J. C. H. Blom, 'Nederland onder Duitse bezetting 10 mei 1940–5 mei 1945', D. P. Blok et al. (eds) *Algemene Geschiedenis der Nederlanden*, vol. 15 (Haarlem: Fibula van Dishoeck, 1982), 55–94; K. van Berkel, 'Kwaliteit en zuinigheid in het wetenschappelijk onderzoek', K. Schuyt and E. Taverne, *Welvaart in zwart-wit* (Den Haag: SDU Uitgevers, 2000), 331–54.

6 H. A. M. Klemann, *Nederland 1938–1948 Economie en samenleving in jaren van oorlog en bezetting* (Amsterdam: Boom, 2002), chapter 16 'Economie en Samenleving', 565–77.

7 For details about H. B. Dorgelo see *Diskus, officieel orgaan van het Eindhovens studentencorps* (Eindhoven: ESC, 1961) (special issue completely filled with obituaries).

8 J. van Gorkom, *'"Es war eine Manie" Een elektronenmikroskoop voor biologen'*, PhD diss. Utrecht University, 1988, 74 (my translation).

9 E. Ruska, *The early development of electron lenses and electron microscopy* (Stuttgart: Hirzel Verlag, 1980), 24–31; T. Mulvey, 'The history of the electron microscope', S. Bradbury and G. L'E. Turner (eds), *Historical aspects of microscopy* (Cambridge: W. Heffer & Sons, 1967), 201–26.

10 Several contributions in P. W. Hawkes (ed.), 'The beginnings of electron microscopy', *Advances in Electronics and Electron Physics*, suppl. 16 (1985) deal with the early research groups in electron microscopy; F. Müller, 'The birth of a modern instrument and its development during World War II: Electron microscopy in Germany from the 1930s to 1945', this volume.

11 Among other publications: W. G. Burgers and J. J. A. Ploos van Amstel, 'Electron optical observation of metal surfaces', *Physica* 4 (1937), 5–14.

12 *Natuur en Techniek* 5 (1935), 321–3; *Nieuwe Rotterdamse Courant*, 25 August 1938.

13 A. C. van Dorsten, W. J. Oosterkamp and J. B. le Poole, 'Een experimenteel electronenmicroscoop voor 400 kilovolt', *Philips Technisch Tijdschrift* 9 (1947), 137–43; *Natuurwetenschappelijk onderzoek in Nederland: een overzicht van hetgeen in de laatste vijf jaren in Nederland verricht is op het gebied der natuurwetenschappen, der medische en der technische wetenschappen*, (Amsterdam: Noord-Hollandsche Uitgevers Maatschappij, 1942), 20.

14 J. B. le Poole, *Optica van licht en electronen* (Delft: Waltman, 1957), 14. In this, his inaugural, lecture le Poole addressed Dorgelo with the following words: 'Hoezeer u erin slaagde, de door u nagestreefde dienende functie te bekleden blijkt wel duidelijk daaruit, dat ik jarenlang ten onrechte gemeend heb, het initiatief tot de electronenmicroscopie in Delft genomen te hebben'.

15 W. de Vries Wzn, 'Waller (Jr), François Gerard (1895–1974)', J. Charité, et al., *Biografisch Woordenboek van Nederland* (Amsterdam: Elsevier, 1985), vol. 2, 610.

16 J. van den Toorn, 'Kluijver, Albert Jan (1888–1956)', J. Charité 1985, vol. 2, 305–7.

17 Philips Company Archives, box 72: *Natlab*, letter from Dorgelo to Holst dated 20 July 1944.

18 Müller 2009.

19 J. G. Smit, et al., *Vijfentwintig jaar Technisch Physische Dienst 1941–1966* (Delft: Centrale Organisatie TNO, 1966); G. A. van de Schootbrugge, *50 Jaar TPD in beweging* (Delft: Technisch Physische Dienst TNO-TU Delft, 1991).

20 W. L. M. Kuijpers, *De Technisch Physische Dienst en de Electronenmicroscopie. Een historische studie naar toepassingsgericht natuurkundig onderzoek bij TNO*. PhD diss. TU Eindhoven, 1987; Smit, 1966, 30–6.

21 B. Elema, *Honderd jaar gistresearch* (Delft: Koninklijke Gist- en Spiritus fabriek N.V., 1970).

22 Copy of this report in Kluyver Archives, box 1990349, Electronenmicroscopie folder.

23 Mulvey and van de Laak-Tijssen 2001, 301.

24 The papers presented at this meeting were published in *Nederlandsch Tijdschrift voor Natuurkunde* 7 (1940) 157–70: H. B. Dorgelo, 'Inleiding over electronenmicroscopen van verschillenden aard'; 171–8: R. Kronig, 'De theoretische grondslagen der elektronenoptica'; 179–91: H. Ruska, 'Onderzoeksmethoden en resultaten der supermicroscopie'.

25 Ruska 1980, 64–5; D. H. Krüger, P. Schneck and H. R. Gelderblom, 'Helmut Ruska and the visualisation of viruses', *The Lancet* 335 (2000), 1713–17 (A German translation of this article can be found on the internet at: http://helmut.ruska.de).

26 Krüger et al., 2000.

27 *Gedenkboek van het verzet der Delftsche studenten en docenten gedurende de jaren 1940–1945* (Delft: DUM, 1947).

28 In some papers (Le Poole 1985; Mulvey and van de Laak-Tijssen 2001) the date is 8 April 1941, in others (Le Poole 1966; Smit 1966, 32–3) the date is 16 April 1941.

29 Le Poole 1985, 394; Mulvey and van de Laak-Tijssen 2001, 305–6.

30 Van Gorkom 1988, 76 (my translation).

31 Van Dorsten 1978; Le Poole 1966; Le Poole 1978; Le Poole 1985; Mulvey and van de Laak 2001.

32 Kluyver Archives, box 1990314, Delftsch Hoogeschoolfonds Electronenmicroscopie folder: draft of a letter from Dorgelo and Kluyver to the Delftsch Hoogeschoolfonds dated 13 May 1942.

33 Van Dorsten 1978, 116.

34 Van Dorsten 1978, 116.

35 Le Poole 1978, 113.

36 Le Poole 1985, 395.

37 Apart from papers in scientific journals by the various scientists involved in the development of their electron microscopes, both Siemens and AEG published more general over-

views of their efforts in 1941: *Die Übermikroskop als Forschingsmittel*, Berlin (Siemens); C. Ramsauer, *10 Jahre Elektronenmikroskopie*, Berlin (AEG). Also, a summing up of the German state of the art in electron microscopy, entitled *Geometrische Elektronenoptik dargestellt unter besonderer Berücksichtigung des Elektronen- und Übermikroskops*, was published in 1942 in the series *Europäische Studienmappe*. It was largely based on the research at AEG. For an historical analysis, see Müller 2009.

38 Le Poole 1985, 396–7.

39 Le Poole 1985, 396.

40 H. J. de Heer, 'Het Delftsche Electronenmicroscoop', *TNO Nieuws* 2 (1947), 51–5.

41 Kluyver Archives, box 1990314, Delftsch Hoogeschoolfonds Electronenmicroscopie folder: letter from the Hoogeschoolfonds to Dorgelo and Kluyver dated 30 July 1942.

42 Le Poole 1985, 394; Kluyver Archives, box 199034, Electronenmicroscopie folder: exchange of letters Kluyver–Quispel September 1942; van Iterson 1996, 273.

43 Philips Company Archives, box 72, Natlab, letter signed by Mulder, dated 17 December 1942.

44 Philips Company Archives, box 72, Natlab, letter from Waller, dated 11 November 1942.

45 Philips Company Archives, box 72, Natlab, copy of a letter from Waller to Philips, dated 18 November 1943.

46 C. J. G. Bakker, *Elektronenmicroscopie in Nederland. Overzicht en evaluatie van 40 jaar speur- en ontwikkelingswerk in een technisch-fysisch specialisme*. Utrecht, FOM rapport nr. 43105, 1977.

47 See many relevant documents in Kluyver Archives box 1990349, Electronenmicroscopie folder.

48 Smit 1966.

49 Philips Company Archives, box 72: *Natlab*, minutes of meeting of the Board dated 24 February 1944.

50 Philips Company Archives, box 72: *Natlab*, letter from Dorgelo to Holst dated 20 July 1944.

51 de Heer 1947.

52 H.J. Reinink, J.H. Bannier (eds), *Nederlandse Organisatie voor Zuiver-Wetenschapelijk onderzoek 1945–1949* (Den Haag, 1950), p. 71.

53 Smit 1966, p. 35.

54 Kluyver Archives box 1990349, Electronenmicroscopie folder, exchange of letters between Kluyver and Quispel , dated 22, 24, 28 and 29 September 1942. Quispel obtained his doctorate only while already working in Delft: *The mutual relations between algae and fungi in lichens*, Ph.D. diss., Groningen, 1943.

55 Kluyver Archives, box 1990314, Delftsch Hoogeschoolfonds folder, letters from Kluyver to the Hoogeschoolfonds dated 29 April and 2 July 1943.

56 A copy of this document was handed to the author by relatives of van Iterson.

57 Autobiographical notes van Iterson.

58 Kluyver Archives, box 1990349, folder: 'Verslag over gistonderzoek m.b.v. het electronenmicroscoop door J.M. van Brakel'.

59 Kluyver Archives, box 1990349, Electronenmicroscopie folder: Reports of activities 1 November 1943 – 31 January 1944 and 1 February 1944 – 30 June 1944.

60 G. H. Jonker, *Vorming en veroudering van verdunde zilverbromidesolen*. PhD diss., Utrecht University, 1943, 66.

61 Kluyver Archives box 1990349, Electronenmicroscopie folder: Minutes of meeting of the Advisory Board of the IEM dated 5 July 1944.

62 Le Poole 1985, 399.

63 Kluyver Archives, box 1990354, electron microscope folder, report entiled: Indrukken van de stand van de electronenmicroscopie in Amerika, verkregen uit enkele Amerikaansche tijdschriften in bezit van de Staatsmijnen.

64 Le Poole 1985, 405.
65 In fact, Manne Siegbahn had built an electron microscope at the Research Institute for Physics in Stockholm in the late 1930s that had been adapted for electron diffraction experiments in 1944/45, but he hardly published anything about this instrument. http://www.msi.se/manne (last accessed 12 May 2007).
66 Smit 1966, 32; Le Poole 1978; Perquin was a pupil of Kluyver.
67 Kluyver Archives, box 1990349, Electronenmicroscopie folder, verslag bespreking 13 October 1945.
68 F. K. Boersma, *Inventing structures for industrial research: a history of the Philips Natlab 1914–1946.* (Amsterdam: Aksant, 2002); Elema 1970; J. Schweppe (ed.), *Research aan het IJ: BPMA 1914- KSLA 1989: de geschiedenis van het 'Lab Amsterdam'* (Amsterdam: Shell Research, 1989).
69 Van Berkel 2000, 332.
70 Boersma 2002, 70.
71 H. Lintsen (ed.), *Research tussen vetkool en zoetstof. Zestig jaar DSM Research 1940–2000* (Zutphen: Uitgeversmaatschappij Walburg Pers, 2000), 24.
72 Schweppe 1989; Elema 1970; Boersma 2002.
73 See M. Burns 'Scientific research in the Second World War: The case for Bacinol, Dutch penicillin', this volume.
74 Le Poole 1985, 406.
75 Müller 2009; N. Rasmussen, *Picture Control, the electron microscope and the transformation of biology in America, 1940–1960* (Stanford: Stanford University Press, 1997).
76 Müller 2009.
77 Rasmussen 1997, 38–9.

6 'Splendid isolation'?

Aviation medicine in World War II

Alexander von Lünen

Collaboration with occupying forces, be it at the political or the scientific level, is always a delicate topic to deal with. Political sentiment furthers a rather one-dimensional point of view, usually depicting the occupiers as greedy and ruthless misanthropes, while collaborators are even worse, since they add some scent of treachery to it. The occupied are regarded as rogue villains when they collaborate or honourable martyrs if they resist doing so. For the occupiers, the stance seems clear: they bag anything they can get their hands on, and are happy if scientists, engineers, or other highly educated people offer their collaboration voluntarily, even more so when they fill in the gaps of the occupiers.

In modern science and technology, with their immanent international networks and standardised methodology, the need for access to the best brains seems particularly pertinent. Especially in a field where scientific knowledge is extremely specialised and specialists are scarce, as in aviation medicine, one would believe the occupiers would have the utmost interest in 'securing' the services of such persons. This paper will show that, despite the low operational level of German aviation medicine, experts from occupied countries were rarely invited or forced to collaborate during World War II. This will be done by taking the wartime experiences of the Dutch aviation physiologist Jacob Jongbloed as a case study. He was well known and well received in the German aviation medicine community in the 1930s, but had to live from hand to mouth (in terms of scientific activity) throughout the war.

As a preliminary, this investigation will outline the set-up of medical research in general, and of high-altitude and aviation physiology in particular. It will then go on to elucidate the particular set-up of aviation medicine in Germany and The Netherlands in the interwar years, to make the case of this paper: despite the lack of experts in Germany in this field, the Germans did not seek 'assistance' from the occupied countries, while at the same time the experts there had to bear with conditions that made further investigations in this field nearly impossible.

Internationality of Physiology

From the very beginning the medical discipline of physiology had an international character. To a large extent this had to do with the status certain medical faculties had, which attracted students from abroad to study there. For example, since American universities produced 'craftsmen' rather than 'scientists',[1] nineteenth-century American students often chose the renowned medical faculties of the Universities of Berlin, Paris or Vienna for their education.[2] As a result of this circulation at the student level, physiology became a very *international* science.

The same goes for high-altitude physiology, although the circulation of personnel here took place at another level. Since it was (and still is) such a specialised field that hardly any country had more than a few experts in its ranks, its development relied on international collaboration. Researchers *had* to go abroad to find like-minded individuals they could discuss their work with, and in order to finance expensive expeditions they chose cooperation rather than competition.

While balloons were occasionally used for such medical studies, because of the restricted space of balloon gondolas, expeditions to mountaintops were arranged to study the effects of high altitude on human physiology.[3]

One such example was the Anglo-German expedition to the Alta Vista summit on Tenerife in 1910, led by Nathan Zuntz (1847–1920), a physiologist from Berlin, together with Hermann von Schrötter (1870–1928) from Vienna, Joseph Barcroft (1872–1947) from Cambridge, and Claude Gordon Douglas (1882–1963) from Oxford.[4]

In 1911, Douglas and his mentor John Scott Haldane (1860–1936), also from Oxford, together with Yandell Henderson (1873–1944) from Yale University and Edward Christian Schneider (1874–1954) from the University of Colorado Springs (a former student of Henderson's) went up to Pike's Peak in the Rocky Mountains (USA) to study human acclimatisation to high altitude. The voluminous report published afterwards supplied the scientific community with a great deal of usable data and the expedition gained broad recognition for its achievements.[5]

These two examples illustrate the immanent internationality in physiology in general and in high-altitude physiology in particular. The negative effect of being cut off from the international scientific community can best be seen in the case of Germany after World War I.

Aviation Medicine in Germany

International cooperation of the kind mentioned above usually took place at a more personal level, that is, upon the initiative of individuals rather than organisations. It blossomed before the First World War, but came to a halt during and after it, at least in continental Europe.[6]

The situation for German aviation medicine in the interwar years was, to put it mildly, miserable. First of all, German science was isolated after World War I. In October 1914, ninety-three eminent German scientists had published a pamphlet *An die Kulturwelt*, in which they denied Germany's responsibility for the outbreak of the war, that Germany would have breached Belgian neutrality and that war crimes had been committed in Belgium. The international scientific community was upset and despised their German colleagues for years to come.[7]

After the war not only did the Treaty of Versailles (ToV) prohibit a German Air Force, it also cancelled all international treaties (except on questions of trade and traffic), including scientific exchange (paragraphs 282ff.). Both restrictions meant a serious blow for the growing field of aviation medicine. Furthermore, with the death of Nathan Zuntz in 1920 German physiology lost its most important champion.

As a result, aviation medicine in Germany in the 1920s was practically non-existent. One of the two exceptions was a clinic for aviation medicine, founded in Hamburg in 1927, although it failed to gain an international reputation.[8] The other was the decision of the Kaiser-Wilhelm-Gesellschaft, predecessor of the Max-Planck-Gesellschaft, in 1930 to participate in the construction of a research laboratory on the Jungfraujoch Mountain in the Swiss Alps to gain access to an international research environment.[9]

While the USA had people such as Henderson, Schneider, Louis Bauer (1888–1964) or Harry Armstrong (1899–1983)[10] and Britain had eminent figures such as Barcroft, Haldane or Leonard Hill (1866–1952),[11] in Germany a whole generation of such specialists was lost owing to the restrictions of the ToV, the rejection from international scientists in the aftermath of World War I and a general lack of resources for scientific research. Attempts to lift the ban on German science, most notably by the Rockefeller Foundation (RF), only partially improved the situation.[12]

On the other hand, when the Nazis came to power in 1933, the 'conflict of the generations at the German universities'[13] ought to have helped the aviation physiologists. Older, established professors were driven out of the universities because of their racial background or their political stance and they were replaced with younger, politically desirable persons. Most German aviators (those interested in aviation medicine were usually aviators themselves) were rather right-wing and had a good standing with the new government. Furthermore, now that a government with pronounced ambitions to re-establish an Air Force ruled the country, physiologists could have hoped for better conditions.

The Nazis, however, did not have a master plan for the organisation of science in mind, other than erecting a 'German' science by eradicating what they considered 'Un-German' science.[14] Aviation medicine was not regarded as vital and played only a minor role in the early Nazi years.[15] Nevertheless, the 1930s saw a break in the isolation of German science, also in the case of aviation medicine. The Germans started to publish the journal *Luftfahrtmedizin* in 1937, at a time when no other European country had a journal specifically dedicated to aviation medicine.[16] They invited researchers from abroad and held numerous conferences on all aspects of aviation, including medicine. One often-seen guest at these conferences and in German journals was the Dutch physiologist Jacob Jongbloed (1895–1974), who gained a good reputation among his international colleagues.

Jacob Jongbloed

Jongbloed proved that a big budget and extensive research facilities were not absolutely necessary to develop the field of aviation medicine. Having been drafted into the infantry on the outbreak of World War I, he soon transferred to the Royal Dutch Air Force, receiving his pilot's licence in 1916 and becoming an officer of the reserve the same year. He remained with the Flying Corps (*Luchtvaartafdeling*) at Soesterberg Air Force Base, where he worked as a flight instructor from 1919 to 1923. During that time he was also a member of the stunt flying team *De Vijf Vingers*, which became renowned for its formation flying skills (Figure 6.1). In 1922, he started to study medicine at Utrecht University (UU), finishing with an MD in 1927. From 1927 to 1931 he served as flight surgeon at Soesterberg, while working on his PhD thesis in aviation medicine, which he completed in 1929.[17] His supervisor was A. K. M. Noyons, professor of physiology at UU. The experiments for his PhD were pursued at Soesterberg, where a decompression chamber was available. His fellow pilot comrades served as test subjects for his investigations into *hypoxia* (deprivation of oxygen in the inspired air due to a decreased barometric pressure) and the effects of acceleration on the human body. Laboratory experiments were done at the Physiological Laboratory of UU.

His thesis gave a summary of the state of the art in the field, utilising the international standard literature of Bauer, Barcroft and Haldane,[18] for example, and it also included results from Jongbloed's own experiments. For instance, he could exemplify the effects of low barometric pressure and high acceleration on the human body. He also discussed the possibility of decompression sickness in aviation.[19] Other researchers, Schrötter or Henderson[20] among them, had already mentioned the theoretical possibility of the *bends* in aviation, but disregarded it

Figure 6.1 A photograph of *De Vijf Vingers*, with two assistants, at Soesterberg, c. 1921. Jongbloed is on the far left. Courtesy of Erno Eskens.

since they considered it not likely ever to occur with contemporary aircraft. Jongbloed, the vigorous pilot, stated in his thesis:[21]

> Both are surely record-flights [current speed and height records], but what is today only possible with a specially constructed aircraft, will be a common possibility in a few years' time.

Jongbloed's research funds remained limited. In 1931, he attained the post of *conservator* at the Physiological Laboratory at UU, adding the post of a *privaat-docent* (lecturer) of *aëro-physiologie* (aviation physiology) to his portfolio in 1934.[22] In 1936, he became adviser to the Ministry of Transport on aviation and aviation medicine (*Raad voor luchtvaart en luchtvaartgeneeskundige van het ministerie van Verkeer en Waterstaat*).

Jongbloed's work was to a large degree restricted to teaching. Genuine research was severely hindered, since he had no direct access to laboratories or funds. Only his former supervisor Noyons granted him occasional access to the labs of the physiological department to conduct research.[23] Jongbloed nonetheless continued to publish papers on the topic of aviation medicine in scientific journals.[24] When Noyons died in the summer of 1941, his position became available, and after some deliberation Jongbloed became professor of physiology in April 1942. His inaugural lecture dealt with general human physiology, the main topic of the professorship.[25] Upon his assignment as a professor, the OKW Ministry (see note 22), as a formal act, withdrew the licences as lecturer for aviation physiology and conservator of the physiological laboratory, since professors were not allowed to hold multiple posts.[26]

The situation ought to have been quite convenient for Jongbloed, now that he had attained a position that gave him full access to laboratories and resources. His research on aviation physiology, hindered by his insufficient funds as a lecturer, should have literally taken off now. Unfortunately, these were times of war, and his position as professor (which he held until he retired in 1965) was equipped with rather small funds. Even though his work had been so well received by German aviation physiologists in the 1930s, his career in this field came to a halt.

Aviation Medicine during World War II

In general, German research activities at the beginning of the war remained comparatively normal. Only after it became apparent to the Germans that the *Blitzkrieg* was no longer successful did they intensify their research activities.[27] Aviation medicine, however, in contrast to research in other branches of aviation (such as aerodynamics or engineering), was not regarded as vital for the war effort.[28]

Convinced of their own position, the aviation physiologists tried to attract the attention of the regime to their research. Hermann Göring himself had concentrated research and development in the area of aviation in the hands of the RLM (see note 28), trying to keep other institutions such as the National Research Council (*Reichsforschungsrat*) out.[29] Hubertus Strughold (1898–1986), director of the Department for Aviation Medicine in the RLM, could therefore easily monopolise and control aero-medical research in Germany. Even though there were a good dozen research institutions and universities that undertook studies in this field during the war, Strughold, as editor of the scientific journal *Luftfahrtmedizin* and coordinator of all research activities in aviation medicine, had *de facto* orchestrated the whole field. Together with his friend Siegfried Ruff (1907–1989), director of the Department for Aviation Medicine of the German Experimental Aeronautical Institution (*Deutsche Versuchsanstalt für Luftfahrt*, DVL) in Berlin, he was in full control of the activities in aviation medicine in Germany – before, during and after the war!

Together, Strughold and Ruff not only wrote the most important German textbook of their time in the field,[30] they also authored the largest number of research reports and papers in Germany in the Nazi era. A revealing picture emerges when the ZWB-index is taken into consideration. This index was compiled under the auspices of the Technical Director of the RLM, the *Zentralstelle für Wissenschaftliches Berichtswesen in der Luftfahrtforschung* (ZWB), and concerned research reports from German institutions in the fields relevant to aviation.[31] Extracting the reports on aviation medicine from it and analysing the author and topic distribution, it becomes clear how far Ruff's and Strughold's influence went. Ruff (33 per cent) and Strughold (11 per cent) not only themselves authored a great number of reports, but also supervised the publication of reports done by their assistants. Examples are Strughold's assistants Clamann (10 per cent) and Opitz (3 per cent) and Ruff's assistants Romberg (8 per cent), Schütze (5 per cent) and Wiesehöfer (5 per cent). In short, Ruff, Strughold and their affiliates published the great majority of the reports.[32]

The reason for so many reports authored by Ruff's and Strughold's assistants becomes apparent when looking at the drafting policy of the German Armed Forces: research considered important for the war effort was given priority, and researchers in such fields were not drafted for military service. To label something as 'important' usually required a letter of recommendation from a scientific director, and Strughold wrote a number of them.[33] Therefore, in order to prevent researchers being drafted for military service, it was important to keep all the 'important' research in their own hands. Hence, German aviation physiologists were not too keen to have competition from occupied countries.[34]

There was another reason why the German physiologists did everything to secure their monopoly on medical research in aviation. Physiologists doing research on aviation were in a fairly comfortable position, since a whole generation of aviation physiologists was missing in Germany owing to the decline of the discipline in the 1920s. Young physiologists now had a good chance of pushing into this vacant field, without struggling with their predecessors. The prospect of making a career in the science of aviation medicine therefore contributed to the self-selected isolation of German aviation physiologists during World War II.

Next, the pronounced *habitus* of the Nazi ideology, of being a *superior race*, in German science in the Nazi era affected the stance towards foreign scientists. The mentality of German aviation physiologists, being the confident aviators most of them were, was dominated by a feeling of *sublimity*. The widespread nationalistic, or even national-socialist, sentiment furthered the exclusion of foreign scientists.

These reasons, together with the scarcity of resources, led to the circumstance that no aviation physiologists from occupied countries were asked to work in or for Germany. In France, for example, neither the eminent physiologist Paul Garsaux (1882–1970), director of a clinic for aviation medicine at Le Bourget airport until 1940, nor George Goett, director of the medical service of the French Air Ministry (*Service de santé du secrétariat d'État à l'aviation*) from September 1940 on, were asked to join. Only a small number of flight surgeons were sent to Germany in exchange for French prisoners, but none of them participated in scientific research. Also, laboratory equipment was bagged and brought to Germany.[35]

The question whether foreign physiologists collaborated or not can therefore easily be disposed of: no one asked them to collaborate in the first place. German aviation physiologists were happy to have their turf and defended it vigorously. So we can speak of a *mutual seclusion*, in which both sides (scientists in occupied countries versus scientists in Germany) had no expressed desire to cooperate.

Splendid Isolation?

In addition to the damage caused by the self-chosen isolation, the restrictions imposed by foreign colleagues inflicted a tremendous decline in terms of scientific exchange, and since the international cooperation was so vital for the discipline (as outlined above), German aviation medicine could not profit from research made in the USA or the UK. The CIOS Report from the summer of 1945 states:[36]

> Although numerous observations were divulged with enthusiasm nevertheless most of them had long since been made and applied to the Allies. It appears therefore that German investigators suffered tremendously from their isolation. Knowledge and developments apropos aviation medicine were not coming to them from outside their sphere of domination.

And this 'sphere of domination' was guarded with great envy. To sustain their elevated status, German aviation physiologists boasted about their achievements. A look at the ZWB-index, however, reveals a somewhat different picture. Only half of the reports dealt with physiological issues, the other half were on human-factor-engineering, accident statistics, proceedings, etc. Of these physiological reports, a good bulk was on *applied* physiology. Some of the inquiries into *pure* physiology were of a rather questionable nature, so the stack of scientifically credible reports is actually rather slim.[37]

Jongbloed and the War

Jongbloed's activities in World War II, on the other hand, are somewhat unclear. His biographers are silent about his role in the time of occupation. Judging from his publications, he did not conduct any further studies in aviation physiology. As far as this author was able to reconstruct, his only publications during the war years were the inaugural lecture mentioned above and an obituary on Noyons in a Dutch newspaper and in the yearbook of the Physiological Laboratory.[38] His scientific activities were apparently restricted to teaching courses in physiology at Utrecht University.

Judging from a transcript of his notebook, Jongbloed was not very keen to keep up 'business as usual' in times of repression. The university was closed for a time during the strikes in April and May of 1943 (*April-Mei-Staking*),[39] owing to subversive activities by students (buildings of the university were set on fire, illegal postings were made). After the riots were subdued by German forces, the university was scheduled to re-open on 1 June, and the professors agreed to 'slowly begin' their teaching again. Jongbloed, however, and another professor refused to take up their lectures again 'under the given circumstances', which led to a controversy. Other professors feared that this might provoke reactions from the Germans. One day later, Jongbloed agreed to comply with the majority's decision 'for the sake of unity'.[40]

This action indicates that Jongbloed was critical of the German occupation and the transcripts vindicate this conclusion. Considering the numerous distinctions given to him after the war,[41] it is at least unlikely that his companions deemed him a 'collaborator'.

Conclusions

While in other fields of research, such as aerodynamics, the Germans were eager to recruit skilled personnel from foreign countries, this was apparently not the case in aviation medicine. This is a somewhat surprising situation, since one would have expected the Germans to be interested in appropriating foreign knowledge and experts in times of war. As has been shown, this abstract political stance found little support among the scientists themselves, who, rather, benefited from isolation. Göring's doctrine of the superior *Luftwaffe* also seems to have played a paradoxical role,[42] leading to the sentiment of superiority of German aviation medicine. As already mentioned, Strughold and Ruff retained their positions after the war. Strughold came to the USA through Operation Paperclip, while Ruff remained director of the medical department of the DVL. One of Strughold's first jobs was to edit a two-volume work on German aviation medicine[43] and it should not be too surprising that it bragged about German achievements.

Judging the scientific quality of German aviation medicine is not the issue here. This author's task has been to question the popular picture of forced collaboration and to show that in times of war, scientific research can be hindered to such a degree that ordinary mechanisms of scientific exchange come to an almost complete halt, and that such a breakdown of information exchange may well be supported by the scientists themselves.

Acknowledgements

I would like to express my gratitude to Mr Marten Tromp of the Archive of the University of Utrecht, Mrs Annemiek de Jong and Mr Frans van den Hoven from the University Museum Utrecht and G. J. Röhner from *Het Utrechts Archief* for their help and assistance in tracking down relevant documents and information on Jacob Jongbloed in their respective institutions.

Thanks also to Erno Eskens (grandson of Captain Bakkenes, former member of *De Vijf Vingers*) for his kind permission to use the photograph shown in Figure 6.1 (Bakkenes is on the far right).

Notes

1 R.L. Numbers and J. Harley Warner, 'The Maturation of American Medical Science', N. Reingold and M. Rothenberg (eds) *Scientific Colonialism. Cross-Cultural Comparison* (Washington DC: Smithsonian Institution, 1987), 195.

2 J. Harley Warner, 'The Fall and Rise of Professional Mystery. Epistemology, Authority and the Emergence of Laboratory Medicine in Nineteenth-Century America', A. Cunningham (ed.), *The Laboratory Revolution in Medicine* (Cambridge: Cambridge University Press, 1992), 110–41.

3 For a history of medical high-altitude studies, see J. B. West, *High Life: A History of High-Altitude Physiology and Medicine* (New York and Oxford: Oxford University Press, 1998).

4 Zuntz and Schrötter became the most influential figures in high-altitude physiology before World War I in German-speaking countries, publishing numerous accounts on mountain and balloon expeditions, and the first monographs in aviation medicine: N. Zuntz, *Zur Physiologie und Hygiene der Luftfahrt* (Berlin: Springer, 1912); H. von Schrötter, *Hygiene der Aeronautik und Aviatik* (Wien: Wilhelm Brauer K.U.K. Hof- und Universitätsbuchhändler, 1912).

5 Cf. C. G. Douglas et al., 'Physiological Observations Made on Pike's Peak, Colorado, with Special Reference to Adaptation to Low Barometric Pressures', *Philosophical Transactions of the Royal Society, Series B* 203 (1913), 185–381. Furthermore, Henderson and Schneider learned much from Haldane on this expedition. In 1917, when the US Army installed the US Army Air Corps and with it its *Medical Research Board* (MRB), Henderson became the director of the MRB, and Schneider was made the director of the physiological section of the MRB. Cf. W. H. Wilmer, 'The Early Development of Aviation Medicine in the United States', *Military Surgeon* 77.3 (1935), 115–35. The four physiologists remained good friends and continually exchanged their ideas in the years to come.

6 The Anglo-American exchange remained comparably good, owing to personal relations between researchers. In France, not much support was given to high-altitude research, since the Ministry of Defence did not see any value in it. Support therefore had to be requested from industry by scientists. Cf. P. Garsaux, *Histoire anecdotique de la médecine de l'air* (Paris: Scorpion, 1963).

7 U. Marsch, *Notgemeinschaft der Deutschen Wissenschaft. Gründung und frühe Geschichte 1920–1925* (Frankfurt/M.: Peter Lang GmbH, 1994), 36.

8 Cf. V. Harsch, *Das Institut für Luftfahrtmedizin in Hamburg-Eppendorf (1927–1945)* (Neubrandenburg: Rethra, 2003).

9 A. Neumann, 'Die Luftfahrtmedizin von der Weimarer Republik bis zur frühen Bundesrepublik', H. Trischler and K.-U. Schrogl (eds), *Ein Jahrhundert im Flug. Luft- und Raumfahrtforschung in Deutschland 1907–2007* (Frankfurt: Campus, 2007), 139.

10 A. Noë, 'Medical principle and aeronautical practice: American aviation medicine to World War II', PhD diss., University of Delaware, 1989.

11 T. M. Gibson and M. H. Harrison, *Into thin air. A History of Aviation Medicine in the RAF* (London: Robert Hale, 1984).

12 The RF tried to loosen the grip of isolation on German scientists by providing exchange programmes. Marsch 1994, 51. One of those who profited from this programme was physiologist Hubertus Strughold, who later became the dominant figure in German aviation medicine (see pp.100–2). From 1929 to 1931 Strughold spent time at American universities on an RF fellowship, meeting aviation physiologists from the States. For a hagiographic account of Strughold, see V. Harsch, *Leben, Werk und Zeit des Physiologen Hubertus Strughold (1898–1986)* (Neubrandenburg: Rethra, 2004). This book is not very critical of Strughold's role in the Nazi era, neither is M. Phillips Mackowski, *Testing the Limits. Aviation Medicine and the Origins of Manned Space Flights* (College Station: Texas A&M University Press, 2006).

13 M. Grüttner, 'Wissenschaft', W. Benz, H. Graml, and H. Weiß (eds), *Enzyklopädie des Nationalsozialismus* (München: dtv, 1997), 135.

14 Grüttner 1997, 136. Through this policy, pioneers such as Adolf Loewy (1862–1936) and Zuntz were not commemorated by Strughold and others, since both were Jews. Most aviation physiologists held pilot's licences themselves (not only in Germany or The Netherlands), with the aviation community in Germany in the interwar years being quite right-wing or '*völkisch*'. Most of the aviation physiologists and flight surgeons in Germany in the 1930s were party members, and 'older' physiologists such as Strughold or Hermann Rein had a pronounced nationalistic, right-wing attitude before the Nazi era. Cf. K. H. Roth, 'Strukturen, Paradigmen und Mentalitäten in der Luftfahrtmedizinischen Forschung des "Dritten Reichs" 1933 bis 1941: Der Weg ins Konzentrationslager Dachau', *Zeitschrift für Sozialgeschichte* 2.15 (2000), 58.

15 Neumann 2007, 140.

16 In the USA, the *Journal of Aviation Medicine* was published from 1930 on by the *Aero Medical Association*, which was founded in 1929, both upon the initiative of Louis Bauer. Cf. Noë 1989.

17 See J. Jongbloed, *Bijdrage tot de physiologie der vliegers op grote hoogten* (Utrecht: Bruna, 1929). Biographical information was taken from various paper clippings upon his 70th birthday in 1965, and his death in 1974. Document collections in the *Utrecht University Museum* (UUM), Box 102, Folder 0285.9443.2. There is also a manuscript of an obituary/biography authored by his former student Peter Kylstra (who published numerous of the biographies/obituaries mentioned) in UUM, 102/0285.9443.1.

18 L. Hopewell Bauer, *Aviation Medicine* (Baltimore: Williams & Williams, 1926); J. Barcroft, *The Respiratory Function of the Blood. Vol. 1: Lessons from High Altitudes* (Cambridge: Cambridge University Press, 1925); J. S. Haldane, *Respiration* (New Haven: Yale University Press, 1922).

19 Also often referred to as *caisson sickness* or the *bends*, due to the bent posture of those who writhed in pain from it. Cf. J. L. Phillips, *The Bends – Compressed Air in the History of Science, Diving and Engineering* (New Haven: Yale University Press, 1998).

20 H. von Schrötter, 'Über Höhenkrankheit mit besonderer Berücksichtigung der Verhältnisse im Luftballon', *Protokoll über die vom 20. bis 25. Mai 1902 zu Berlin abgehaltene dritte Versammlung der Internationalen Kommission für wissenschaftliche Luftschiffahrt* (Strassburg: DuMont-Schauberg, 1903), 102–29; Y. Henderson, 'The Physiology of the Aviator', *Science* 49.1271 (1919), 431–41.

21 Jongbloed 1929, 2, translation mine. As a matter of fact, *decompression sickness* in aviation did not become an issue before the introduction of propulsion jet engines in the 1950s.

22 This post was established at UU in 1922 at the request of the Ministry of Education, Art, and Science (*Ministrie van Onderwijs, Kunsten en Wetenschapen*; OKW Ministry).

In 1920 the Air Force base at Soesterberg had established the air medical service for screening pilot candidates and treating pilots. To benefit from new scientific research, the Ministry of War requested the cooperation with medical faculties. H. Zwaardemaker (1857–1930), then professor of physiology at UU, and his former student P. M. van Wullften Palthe (1891–1976), also air force officer of the reserve, who wrote his PhD dissertation on aviation psychology (P. M. van Wulfften Palthe, 'Zintuigelijke en psychische functies tijdens het vliegen', PhD diss., Utrecht: Rijksuniversiteit te Utrecht, Faculteit voor Geneeskunde, 1921), were nominated as likely candidates. In 1922, Wullften Palthe was given the newly created position of lecturer of aviation physiology at UU. See: Letter from the OKW Ministry to the *Curatoren der Rijksuniversiteit te Utrecht*, 26 January 1920, *Het Utrechts Archief* (HUA), Toegang 59, Inv.-Nr. 603; Letter from the Medical Faculty (UU) to the Curators (UU), 29 July 1922, HUA, 59/1134. For Jongbloed's receiving of the position, see: Letter from the Medical Faculty (UU) to the Curators (UU), 18 December 1933, Letter from the OKW Ministry to the Curators (UU), 24 January 1934, Letter from the Medical Faculty to Jongbloed, 2 February 1934, all in HUA, 59/1134.

23 In 1935, he managed to obtain two grants from the Donders-Foundation (*Donders-Fond*), which led him to the laboratory of renowned Belgian neurophysiologist Prof. C. Heymans in Ghent, and several medical institutes (Pavlov's among them) in the USSR. On both trips he ventured into neurophysiology and learned techniques of autopsy new to him; in Ghent he also lectured on the effects of acceleration on the nervous system in aviation. Cf. HUA, 59/3456.

24 Many of them in German journals, e.g. J. Jongbloed, 'Über das psychische Verhalten während kurzen Aufenthaltes auf 5000m Höhe', *Klinische Wochenschrift* 14.44 (1935), 1564–8; J. Jongbloed and A. K. Noyons, 'Der Einfluß von Beschleunigungen auf den Kreislaufapparat', *Pflügers Archiv* 233.1 (1934), 67–97; J. Jongbloed and A. J. H. Wildschut, 'Kohlensäure- und Sauerstoffzusatz bei Höhenatmung', *Luftfahrtmedizin* 3.1 (1938), 8–11; through which he achieved a good reputation in Germany among aviation physiologists. In 1937, he gave a paper at the First International Conference of Aviation Medicine in Berlin. His invitation by the German *Lilienthal-Society* to their meeting in 1939 was rescinded when the meeting was cancelled owing to the outbreak of World War II.

25 Cf. J. Jongbloed, *Beschouwingen over het onderwijs in de physiologie: voornamelijk te Utrecht* (Utrecht: Bruna, 1942).

26 Cf. letters between Jongbloed and Curators between April and May 1942, HUA, 59/1134.

27 Grüttner 1997, 149.

28 The *Reichsluftfahrtministerium* (RLM, German Air Ministry) was convinced that high-altitude bombers and high-altitude fighters could even out the odds in air warfare. These projects did not start before 1941, and construction issues with Junkers, Messerschmitt, BMW and Daimler were at the top of the agenda. Aviation medicine played virtually no role in the discussions in the RLM. Cf. minutes of the meetings of the Technical Director of the Air Ministry (*Generalluftzeugmeister*), E. Milch, in the years 1941–1943, *Bundesarchiv-Militärarchiv* (BA-MA) Freiburg (Germany), RL 3, Boxes 21, 24, 37, 60.

29 This, obviously, included aviation medicine. Neumann 2007, 144.

30 S. Ruff and H. Strughold, *Grundriß der Luftfahrtmedizin* (Leipzig: Barth, 1939), which was re-edited 1944 and 1957.

31 See H. Trischler, *Luft- und Raumfahrtforschung in Deutschland 1900–1970. Politische Geschichte einer Wissenschaft* (Frankfurt a. M.: Campus Verlag, 1992), for a structure of aviation research in Germany. For a contemporary account on the ZWB, see *Report Index on German Aeronautical Research Documents*. Report No. 948. FIAT, 1945. A good number of the ZWB-reports were also re-published in *Luftfahrtmedizin*.

32 Fifteen per cent of the reports were published by Theodor Benzinger (1905–1999), Director of the Medical Section of the Testing Facility of the Air Force (*Erprobungsstelle der Luftwaffe*) at Rechlin from 1937–1944, and another 8 per cent by Hermann Rein (1898–1953) from the University of Göttingen, who was closely affiliated with Strughold's department and was a co-editor of *Luftfahrtmedizin.*

33 Neumann 2007, 144. Another example would be Adolf Butenandt, biochemist and friend of Theodor Benzinger, who used the latter's standing with Erich Hippke to negotiate an 'appropriate' assignment of two of his assistants, i.e. far from the front. Cf. B. Gausemeier, 'An der Heimatfront. "Kriegswichtige" Forschungen am Kaiser-Wilhelm- Institut für Biochemie', W. Schieder and A. Trunk (eds), *Adolf Butenandt und die Kaiser-Wilhelm-Gesellschaft. Wissenschaft, Industrie und Politik im 'Dritten Reich'* (Göttingen: Wallstein, 2004), 139. That the wrong standing could prove fatal is exemplified by the case of Heinrich Lottig (1900–1941), former director of the clinic for aviation medicine in Hamburg. This clinic became a department of Strughold's institute in 1939 and its significance and autonomy was therefore severely restricted. Lottig, once an influential figure in German aviation medicine, was consequently assigned to a paratrooper unit as a flight surgeon. In May 1941 he was killed in action when serving with the paratroopers in Crete. Cf. L. Brauer, 'Dem Chef des Sanitätsamtes des NS-Fliegerkorps Prof. Dr. Heinrich Lottig zum Gedächtnis', *Luftfahrtmedizin* 6.1 (1942), 4–6.

34 One thing that might exemplify this tendency is the fact that research results of one project were often published redundantly, i.e. the physiologists concerned published several reports and papers under their respective names, simply detailing the same results in varying forms. For example, at the end of the 1930s the DVL undertook a study of the effects of high-speed flying. Siegfried Ruff and his assistants in these experiments, Otto Gauer and Hans Wiesehöfer, all published numerous articles, reports and scientific papers, amounting to more than half a dozen publications, all on the same experiments and results.

35 J. Timbal, 'L'Œuvre Aéronautique du Médecin Général George Goett pendant la Deuxième Guerre Mondiale', *Médecine et Armées* 34.3 (2006), 281.

36 A. J. Vorwald, *Aviation medicine, general medicine, veterinary medicine, chemical warfare*, Report No. XXVIII-59, CIOS, 1945, 5. CIOS (*Combined Intelligence Objective Sub-Committee*) was one of the Allied commissions that surveyed German science and industry between 1944 and 1947, both to de-Nazify it and to look for promising research and development. Other bodies were the *British Intelligence Objectives Sub-Committee* (BIOS), the *Field Information Agency, Technical* (FIAT) and the *Joint Intelligence Objectives Agency* (JIOA).

37 Aviation medicine is considered *applied* medicine by the medical profession and therefore hardly ever gets as much credit as work in other fields. To publish one's research results to a wider audience, one would have to signify the importance of aviation medical research for general physiology. Strughold and Ruff, as the dominant executives in aviation medicine in Germany between 1934 and 1945, could not deliver such signifying aspects. Consequently, no papers from either one were published in the eminent physiological journal *Pflügers Archiv* in that time (Ruff published only one in 1973), while Jongbloed (with Noyons) published three between 1934 and 1938. Other German aviation physiologists such as Benzinger also published next to nothing in it (one paper in 1939).

38 There was one paper in a French journal, and one in the German journal *Luftwissen* in 1940, but both were authored and accepted in 1939 (before the war) and printed and published one year later.

39 Cf. W. Warmbrunn, *The Dutch under German occupation: 1940–1945* (London: Oxford University Press, 1963).

40 Transcript of Jongbloed's notebook (*Aantekeningenboekje*) for the years 1942–1943, UUM, 102/0285.9443.1.

41 He became director of the newly installed Institute of Aviation Medicine at Soesterberg in 1952; became a member of the Royal Dutch Academy of Sciences (*Koninlijke Nederlandse academie van wetenschappen*) in 1949; *Rector Magnificus* of Utrecht University 1958–1959; received the *Snijder-Medal* (*Medaille van de Stichting 'Het Generaal Snijders-Fonds'*, also called *Snijders-Penning*) for his contributions to aviation in 1965; and became Knight of the Order of the Dutch Lion (*Ridder in de Orde van de Nederlands Leeuw*) and Officer in the Order of Oranje-Nassau (*Officier in de Orde van Oranje-Nassau*). See paper clippings in UUM, 102/0285.9443.2.

42 Leading, for example, to the abolition of psychological screening of pilot candidates in 1942, since a 'rational application of accepted principles was made impossible by the official policy of refusing to admit the existence of neurosis, or its precursors, in the Herrenvolk'. D. Williams, *Neuropsychiatric Organizations in the German Air Force*, Report No. XXVI-81. CIOS, 1945, 16.

43 *German Aviation Medicine: World War II*. Prepared under the auspices of the Surgeon General, U.S. Air Force. Dep. of the Air Force. Washington DC: US Government Printing Office, 1950. It is somewhat disturbing to see that historians still cling to this work when evaluating the quality of German aviation medicine.

7 National Socialism, human genetics and eugenics in The Netherlands, 1940–1945

Stephen Snelders

National Socialist warfare would not only lead to the dominance of the 'master race'. In the process of war the race itself would be forged. In this sense racial hygiene and eugenics, pursuing health policies that aimed at the purification of the German people and its racial composition, held a key role within the Nazi bid for dominance and conquest. Not only did the war offer opportunities for pursuing eugenic policies, these policies were considered essential for maintaining and strengthening the fighting powers of the Axis.

It was for this purpose that scientific resources were recruited and alliances made with scientists, not only in Germany itself, but also, during the war, in occupied 'Germanic' countries such as The Netherlands. As will be shown in this paper, the attempt to establish alliances in the Dutch scientific community was not limited to overt sympathisers with Nazism. The methods and aims of racial hygiene did, after all, overlap with those of established sciences such as (human) genetics and physical anthropology. As in Germany, there were in The Netherlands not a few scientists from these disciplines who were to a degree sympathetic to the 'eugenics of the deed' practised in the Third Reich and who saw opportunities to realise their own ideas for healthcare policies under German occupation. This did not, however, necessarily mean that they condoned more extreme policies of anti-Semitism, mass murder or involuntary euthanasia.

The Dutch urologist and National Socialist Constant Croin was president of the *Artsenkamer*, the corporation of physicians of which every doctor was obliged to be a member under the German occupation. In 1942, Croin declared: 'We build our position on the basis of blood and soil, race and heredity: environment can bring talents to full development, but can never add one talent'.[1] The opinion expressed in the second part of this sentence was by no means limited to National Socialists, but was common among Dutch researchers of heredity in genetics, anthropology and medicine. It became discredited in the public sphere because of the Nazi crimes against humanity. In the post-war trials of Dutch National Socialist doctors, biologists and anthropologists their public statements about eugenics and race were held against them, though they were not in themselves sufficient reason for conviction.[2]

It is not surprising, therefore, that the non-Nazi researchers of heredity and eugenicists emphatically tried to dissociate themselves from the ideas that had led to Auschwitz. After the war a victor's history of genetics and eugenics was written, in which the non-Nazi eugenicists had to distance themselves from the 'perversions of Nazism' in order to free their discipline from moral taints.[3] The Nazi eugenicists remained silent.

As usual, with the benefit of hindsight, historical processes prove to have been a little more complex. Developments in The Netherlands show, as argued in this paper, that the final demarcation line between respectable and Nazi genetics and eugenics was the outcome of a process that was produced as much, if not more, by the course of the war as by internal differentiations within the disciplines involved.

Before the War

The complexity of distinguishing between 'respectable science' and 'political aberration' was already evident before the war, in the reception of Nazi Germany's health policies in Dutch geneticist and eugenicist circles. We can see this in the most important forum for eugenic discussions in The Netherlands: *Afkomst en Toekomst* (Ascent and Descent). This journal was founded in 1935 under the name *Erfelijkheid bij de Mens* (Human Heredity) and changed its name in 1937. It had approximately 300 subscribers. All Dutch researchers of genetics and human heredity of any standing published in the journal. M. J. Sirks, since 1937 extraordinary professor of genetics at the University of Groningen, was editor and from 1940 on editor-in-chief. Sirks also fulfilled an important role in the Dutch Eugenic Federation, a federation of all eugenic organisations founded in 1930 with royal approval. Its chairman was G. P. Frets, psychiatrist at the Maasoord asylum in the village of Poortugal. Other important members included the ophthalmologist P. J. Waardenburg, a geneticist of international renown, and the Jewish general practitioner J. Sanders. The geneticists and eugenicists involved with the journal *Afkomst en Toekomst* were very interested in practical applications of their ideas. Some of them, including Sirks, were involved in attempts to examine potential colonists of the Zuider Zee polders for possible hereditary defects.[4]

As is well known, *Afkomst en Toekomst* was published at a time when eugenic ideas were being extensively incorporated into legislation and medical practices of the eastern neighbours in Nazi Germany. The *Gesetz zur Verhütung erbkranken Nachwuchses* led to compulsory sterilisation of potential carriers of hereditary diseases, including feeble-mindedness, schizophrenia, epilepsy and Huntington's chorea. Hereditary Health Courts (*Erbgesundheitsgerichte*) in the Third Reich condemned between 300,000 and 400,000 citizens to compulsory sterilisation, mainly before the outbreak of the war. In 1935, marriages between 'Aryans' and others were made illegal, and evidence of hereditary health was made mandatory for marriage among Aryans (though in practice the situation remained more flexible).[5] Comments on these developments in *Afkomst en Toekomst* varied. Often they were neutral and without criticism. The first publication of the biologist from the University of Groningen, Wouter Ströer, a future member of the Dutch SS, advocated eugenic principles in marriage counselling and legislation, as practised in the new Germany. His co-author was the then editor-in-chief of the journal, J. van Schouwenburgh.[6] Sometimes German authors were themselves given the opportunity to explain the developments. Among them was Ernst Rüdin, one of the scientists behind the German sterilisation law and an international authority as professor of psychiatry in Munich and successor of Kraepelin.[7] But there were critical voices too. Especially the German interpretation of the concept of 'race', a concept that was in itself generally accepted, was problematic to some. Jan Boeke, professor of histology and embryology in Utrecht, commented on the idea of a 'Jewish race': races were not nations, but every nation was composed of different races. He therefore advocated mutual tolerance between the races.[8] Editor-in-chief van Schouwenburg wished to draw a line between eugenics in Germany and anti-Jewish policies. He referred to the difference, made by the moderate Dutch eugenicist and secretary of the Eugenic Federation, A. L. Hagedoorn, of a 'contemplative, complete' eugenics and the practical, consequently one-sided eugenics of the Third Reich. But one cannot say that this German 'eugenics of the deed' was rejected as such.[9] And the demarcation lines concerning 'race' with some leading SS researchers were not that clear-cut either. We will see later how SS doctors and geneticists actually agreed with the theoretical viewpoint of Boeke, who, however, drew different conclusions from it.

SS Geneticist Wouter Ströer

A key figure in linking the pre-war eugenic networks with Nazi eugenics during the war was the aforementioned Wouter Ströer (1907–1979). Ströer had studied biology at the University of Amsterdam. After taking his doctor's degree in 1933 he moved to the anatomical-embryological laboratory of the medical faculty of the University of Groningen, where he was assistant, chief assistant (1934) and prosector (1937). In his experimental embryological research Ströer was especially interested in the heredity of deformations, on which he, for instance, published in the German medical journal *Der Erbarzt*.[10] One of these deformations was harelip, by coincidence also around that time the research topic of the medical doctor dissertation of Josef Mengele.[11] Ströer became more and more interested in human heredity and its application in racial hygiene.[12] When war broke out Ströer held a Rockefeller Fellowship at Yale University in the United States. He hastily broke off his engagements to return to Europe and was back in The Netherlands in time to witness the German invasion.[13]

To Ströer, it seemed that the time had come for the creation of a new, better society, based on the principles of human behaviour as discovered by the researchers of heredity. He took up his old job at the University of Groningen, where the medical faculty became a hotbed of Nazi sympathisers. Ströer was a protégé of the pro-German professor of anatomy and embryology, Herman de Burlet.[14] On 25 July 1940, Ströer became a member of the most important National Socialist party in The Netherlands, the NSB (*Nationaal-Socialistische Beweging*). Before the war this had been prohibited for civil servants, including those working at universities. On 24 October Ströer went one step further and also became a member of the Dutch SS (*Nederlandsche SS*). He was one of the first fifty members of this organisation, founded on 11 September 1940.[15]

Ströer was not the only academic or intellectual who joined the Dutch SS. This is all the more remarkable, since a major part of the activities of SS members was supposed to be devoted to physical exercise, including target practice and sport. An 'SS man' had to sacrifice at least 24 hours of his spare time to the SS every month, 16 hours of which to this physical exercise.[16] The Dutch SS was meant to be an intellectual and military elite of 'political soldiers', propagating the ideas and practices of Himmler and the German SS in the racially affiliated, 'Germanic' Netherlands. Ströer, the talented young researcher of heredity, was meant to play a central role in this project.

Eugenics and National Socialism in the Early Years of the War

Although Ströer went around in his black uniform with the death's head insignia from the autumn of 1940 on, distance between him and the non-Nazi eugenicists widened only slowly. Partly this had to do with the fact that the latter had to find a new political orientation in a new world that seemed to be dominated by the German victors. It further seems probable that the word 'SS' did not mean that much to most Dutchmen at this stage of the war. Most importantly, both the theoretical and practical ideas of the eugenicists were close to those of Ströer and his associates. For example, Ströer's opinion in *Afkomst en Toekomst* in 1939, that not just the interests of the individual should be regarded in the treatment of infertility, and that infertility of persons with hereditary defects should not be treated, was completely in line with the introduction of Sirks and psychiatrist G. W. Kastein to their textbook *Geneeskunde en erfelijkheid* (Medicine and Heredity). Published in 1941, in the midst of the German occupation, the editors of this volume claimed that more attention should be given to the hereditary foundations of public as well as individual hygiene. As individuals, we could consider how to deal with our inherited constitution and its consequences for our personal healthcare; at a social level, we could prevent the birth of children with hereditary defects. The editors further pointed to the increased financial costs of care for the hereditary feeble-minded and insane.[17] In 1941, *Afkomst en Toekomst* regularly published pleas for mandatory medical examination before marriage, and even an advocacy of sterilisation as a humane alternative for segregation.[18] Clearly, these kinds of ideas were not limited to one part of the political spectrum. Kastein was, after all, a communist and member of the armed resistance; he was involved in the murder of the Dutch general and titular commander of the Dutch volunteer's legion in the Waffen-SS Seyffardt, and committed suicide in 1943 after his arrest by the *Sicherheitsdienst*.[19]

On the other hand, this last 1941 volume of *Afkomst en Toekomst* no longer mentioned Ströer as a contributor, and his article on the heredity of morphological characteristics in the textbook *Geneeskunde en erfelijkheid* was published anonymously.[20] It is probable that this was done at his own request, since as a prominent racial biologist of the Dutch SS his name could hardly be in the same table of contents as that of the Jew Sanders (who would not survive the war).[21] But this did not mean that there were no contacts between the SS and non-Nazi geneticists.

Soon after its foundation, the Dutch SS had developed a plan for an institute for anthropology, eugenics and racial hygiene in The Netherlands, along the lines of the *Kaiser-Wilhelm-Institut für Anthropologie, menschliche Erblehre und Eugenik* (KWI-A) in Berlin. The institute should play a central role in the reformation of Dutch healthcare along the lines of the German model. For this, the SS tried to interest the leading Dutch geneticists and eugenicists. The main mover of the plan was the physician Jan Arie van der Hoeven (1912–1998). In 1940, he received his doctor's degree at the University of Amsterdam and became assistant in a sanatorium in the town of Wijk aan Zee. Though Van der Hoeven never became a member of the National Socialist party, the NSB, he was one of the original members (no. 17) of the Dutch SS and from the start a staff member.[22] Van der Hoeven interested Ströer in his plan when the biologist joined the SS. The idea was that doctors and biologists of the Dutch SS would visit German institutions to gain more expertise. Besides Ströer, another National Socialist from the University of Groningen was involved: Aafko Willinge Wittermans (1911–1999), since 1938 assistant there at the psychiatric-neurological clinic as well as leader of the National Socialist students. He would be the expert in the field of mental disorders.[23] At the request of the *Höhere SS- und Polizeiführer* in The Netherlands, Rauter, van der Hoeven would spent a month at the *SS-Rasse- und Siedlungs-*

Figure 7.1 Dutch Nazi doctors of the Medical Front.
Source: National Archive

hauptamt, at the department for hereditary biology (*Erbbiologie).*[24] In February 1941, van der Hoeven also gained the support of the leaders of the KWI-A, Eugen Fischer and Ottmar von Verschuer.[25]

From the beginning, van der Hoeven tried to interest leading Dutch eugenicists in the foundation of an institute for anthropology, eugenics and racial hygiene. He contacted Sirks, the German dermatologist and professor in Leiden H. W. Siemens (author of a textbook on heredity and racial hygiene), and the Catholic gynaecologist, M. A. van Bouwdijk Bastiaanse. Did they reject any contact with the staff of the Dutch SS? Their reaction was more complicated. According to van der Hoeven, Siemens was friendly but rejected the plan. Van Bouwdijk Bastiaanse was at first sympathetic. On a second visit, though, it turned out that he had become a member of the *Nederlandsche Unie* (Dutch Union), a political organisation that tried to create a political alternative to the Nazi party under the German occupation. This ended further cooperation. Sirks was the most interested. Again, according to van der Hoeven, Sirks told him of his sympathy for fascism and his many German friends, but he disapproved of the German invasion and found Jewish influence in The Netherlands of no importance. Sirks liked the idea of an institute, but would not go so far as to support sterilisation legislation.[26] Sirks remained cautious, but did see prospects for his research under the New Order. In July 1942, he requested the National Socialist head of the department for Education for an ordinary chair at the University of Leiden for himself, and an extraordinary chair for Waardenburg.[27]

By that time, the summer of 1942, the execution of the plans of the SS had been delayed. Van der Hoeven had left for the Eastern Front as Waffen SS medical officer after the invasion in the Soviet Union and returned only in May 1942, with frozen feet and dysentery. But now he left active service and started work again on his plans for the research institute.[28] Wittermans had gone to Munich in 1941 to study with Rüdin and returned in 1943. Ströer, who was withheld from service with the Waffen SS on the Eastern Front by Dutch SS leader Henk Feldmeijer, went to the University of Königsberg in 1942 and in March 1943 moved to the KWI-A in Berlin, to be guest assistant of Verschuer.[29]

Nevertheless, van der Hoeven managed to create an office for anthropology and eugenics within the Dutch SS, by then called the Germanic SS in The Netherlands.[30] The methods he intended were those generally accepted among biologists and anthropologists. For example, he tried to get medical colleagues to make a card index system of the medical, hereditary and racial characteristics of all inmates in mental asylums.[31] He further intended to make an inventory of all Indonesian families, with their anthropological data, to establish whether Indonesian blood was to be treated in the same way as Jewish blood; to study the heredity of diabetes; and to found a laboratory for paternity research.[32] As late as November 1942, with the fighting in Stalingrad in full swing, discussions on possible cooperation were held between van der Hoeven and people such as Sirks and the chairman of the Eugenic Federation, Frets, and even again with van Bouwdijk Bastiaanse.[33] But ultimately the war had made the mutual distance too great. By this time cooperation with National Socialists had, with increasing German repression, become much less acceptable in the public view than in the less polarised political climate of 1940. On 23 November 1942, van der Hoeven reported to Ströer that the discussions had failed, and that from now on the SS would do the job on its own.[34]

The SS and Eugenics in the Final Years of the War

Van der Hoeven now started to organise education courses on *erfgezondheid*, hereditary health, for Dutch Nazi doctors, scientists and officials. In cooperation with the 'brains of the Dutch SS', the professor in classical archaeology at the university of Amsterdam, G. A. S. Snijder, an ambitious proposal was sent to the scientific institute of the German SS, the *Ahnenerbe*. It asked for funding for the new eugenic institute. Plans for the institute included the creation of a database with the hereditary biological characteristics of the whole Dutch population. This database, work on which had already been started locally by some SS doctors, would deliver the scientific foundations for future policies of racial hygiene. Furthermore, the institute would work on creating understanding among the Dutch people for the '*von der SS propagierten Auslesungsdanken*', the ideas on biological selection of the SS. An affiliated institute for paternity research, detailing the racial health of future parents, would also be created. Ströer was to be invested with a chair at the University of Groningen.[35]

By that time, Ströer had made a very favourable impression at the KWI-A. Verschuer called him '*eine hervorragende Forscherpersönlichkeit*' ('an excellent researcher') and wanted him as head of a new research department of embryology at his institute.[36] More than 40 years later, when Ströer had died, Dutch researchers claimed that he had had knowledge of the Auschwitz experiments of Josef Mengele, who was said to have sent preparations of organs of murdered prisoners to the Dutchman.[37] Whether Ströer was in the know about Mengele's activities cannot be confirmed or denied. Recent historical research suggests that the members of the KWI-A had their dark suspicions about Mengele, but knew nothing for certain and especially did not want to know anything for certain.[38]

Ströer returned to The Netherlands at the end of 1943. He became extraordinary professor of genetics in Groningen in January 1944. Together with, amongst others, van der Hoeven and Wittermans he edited the journal *Rashygienische Mededeelingen* (Communications on Racial Hygiene). Van der Hoeven himself was made municipal doctor of Amsterdam in May 1944, where, according to one nurse, he never talked about politics at his work, though he did wear the ring with the SS runes.[39]

The Allied invasion of Western Europe shattered the dreams of the SS eugenicists of creating a new racial hygiene in The Netherlands. In September 1944, with the advance of the Allied armies, nearly all Dutch SS men were called to arms in Waffen SS units. Ströer witnessed the Allied landings and the fighting at Arnhem as an SS soldier in the *Landstorm* brigade. After the fall of the Third Reich his scientific career came to an end. His wife poisoned herself and their three children: only one survived.[40] After the war he received a five-year prison sentence.[41] Van der Hoeven also ended up in the *Landstorm* and directed a military hospital in Amerongen. He received a six-year sentence, but was 'sent to work' in the Dutch East Indies, where he had been born.[42]

What Was Specific about Dutch Nazi Eugenics?

As the author has tried to show, the distance between Nazi and non-Nazi geneticists widened only slowly and ideas of cooperation could still be entertained as late as November 1942. It is interesting that up to this point counterfactual developments were still possible. In an alternative world where Nazi Germany had won the war the elements were there for the integration of Dutch genetic and eugenic research in, and its cooperation with, health policies concentrated on hereditary health and even racial hygiene, similar to what had happened with genetics in Nazi Germany itself. But the reason for this is not simply because genetic thought was by definition close to fascism or, to be more precise, National Socialism. Eugenic thought was, on the contrary, clearly something that was shared throughout the political spectrum in The Netherlands, as elsewhere.

This brings us back to the question of what did constitute the essential demarcation line between Nazi and other versions of genetics and eugenics. It was not just the Nazis who wished to come to practical applications of their ideas and research. A large percentage of the non-Nazi eugenicists were in favour of medical examinations before marriage, and examination and biological selection of the Dutch colonists of the Zuider Zee polders. But before 1940 they had no political power base behind them. Nazi eugenicists, on the other hand, did from 1940. Of course, they stood at the radical end of the eugenic spectrum, advocating extreme measures. Nevertheless, SS doctors were as well aware as the non-Nazi eugenicists that their ideas on hereditary defects were not always scientifically robust. But they were of the opinion that they had enough scientific grounds to show the chances of the expression of hereditary illness, regardless of its exact causative mechanisms – for instance, by using the statistical methods of empirical heredity prognosis developed by Ernst Rüdin. This was one of their grounds for supporting forced sterilisation.[43]

One might believe that the real difference between the SS doctors and the non-Nazi geneticists was in their perspective on race. Healthcare was, after all, only meant for the racially pure Germanic people in the view of the SS. Not only did the SS aim at a positive and negative eugenics to strengthen the hereditary composition of the Germanic people, but also the great threat of racial contamination by the Jews, and by the Indonesians from the Dutch colonies, had to be extinguished. After all, as SS doctor and physician at the Rotterdam municipal health service Henk Scalongne claimed, when two different races were mixed the chances of harmonic combinations of hereditary factors were slight.[44] Even here, however, the distance with non-Nazi counterparts was not that wide. Beliefs in the existence of separate races with distinct hereditary characteristics were common, and it was not just Nazis who were negative about racial mixture. Waardenburg advocated limitation of racial cross-breeding and political measures to integrate 'racial bastards' in 'the numerically strongest race of origin'.[45] He did, however, reject the 'exaggerations' of German anti-Semitism. Ultimately, all humans were 'a plastic product of a common original form', and scientists should establish 'quietly and objectively' where racial cross-breeding was inappropriate.[46]

Of course, for SS researchers, racial science was not only a matter of cold scientific objectivity. Haarlem general practitioner G. W. Hylkema explained that cool science now was connected to a racial consciousness, the conviction of people working and fighting miles from academics that survival of their race was at stake.[47] In courses on hereditary healthcare for doctors, it was emphasised by the SS that the idea of an objective, universal science was a fiction. Science was related to culture, and culture to race, in this case the 'Germanic Nordic race'.[48]

At the same time, SS doctors and biologists did consider themselves scientists. They fully shared the view of opponents that no pure races existed, sometimes to the incomprehension of their own political associates. In March 1942, the medical examiner of the Dutch SS, Louis Delbaere, held a lecture for National Socialist doctors and scientists. He explained that races were in continuous development and mutation. No pure, homozygotic races existed. This was reported to the education leader of the Dutch SS, Jan Nachenius, as if Delbaere had denied the existence of races as such.[49] But Nachenius shared Delbaere's opinion. A pure race did not exist, and never had: no, the SS was in the process of creating a *more pure* race.[50]

Nor did SS researchers deny the importance of environment or nurture in shaping phenotypes. In this they were fully in line with the research at the KWI-A, where, since the end of the 1930s, Fischer and Verschuer tried to escape from the dichotomy between the black boxes of genotype and phenotype and wished to study the *Phänogenetik*, what we now might call epigenetics.[51] They were in line as well with pioneers of Dutch human genetics such as M. A. van Herwerden and Waardenburg, who never denied the influence of environment in developing factors present in human beings.[52] Of course, maintained Ströer, one could not escape one's racial hereditary predisposition. Germanic man was bound to his acceptance of fate and feelings of honour.[53] But genetics was not a doctrine of scientific predestination, except in extreme cases of hereditary disease.[54] Dutch Nazi doctors were therefore of the opinion that it was not possible to create strict dividing lines between nature and nurture. Besides eugenic measures, they advocated social hygienic measures as well: improvements in living conditions, better diet and hygiene in work places and the stimulation of sport.[55]

We might conclude that what ultimately divided the Dutch SS doctors and scientists from the non-Nazi geneticists was less the content of their eugenic ideas as such, as the former's linkage to a power base that alienated itself slowly but increasingly from the Dutch population, while at the same time urging more radical stances from its supporters. This isolated the Nazi eugenicists even where they rejected some forms of unscientific thought on race and genetic determinism. The dynamics of shaping demarcation lines between National Socialism and 'democratic' genetics and eugenics were determined more by Nazi Germany's radicalism and its losing the war than by inherent characteristics of the scientific positions involved.

Notes

1 Speech by C. C. A. Croin 12 September 1942. In: Ph. de Vries, *Geschiedenis van het verzet der artsen in Nederland* (Haarlem: H. D. Tjeenk Willink & Zoon, 1949), 300.
2 For example, the proceedings against Dutch Nazi physician Koenraad Keyer. In: National Archive The Hague, Centraal Archief Bijzondere Rechtspleging (CABR) 6232b.
3 Cf. I. Mols, 'Een beladen erfenis. Het raciale vertoog in de sociale wetenschap in Nederland 1930–1950', *Volkseigen. Ras, cultuur en wetenschap in Nederland 1900–1950* (Zutphen: Walburg Pers, 2000), 129–55.
4 J. Noordman, *Om de kwaliteit van het nageslacht. Eugenetica in Nederland 1900–1950* (Nijmegen: Sun, 1989), 109–11, 120–6. For Sanders: G. J. Bremer, *Dr. Jacob Sanders. Een joodse huisarts in het interbellum*, unpublished. Available from the author.
5 Two recent overviews of the historical literature: M. Burleigh, *The Third Reich. A new history* (London: Pan Books, 2001), 345–81; R. J. Evans, *The Third Reich in power 1933–1939* (London: Penguin Books, 2006), 509–610.
6 W. F. H. Ströer and J. Ch. v. Schouwenburg, 'Het kiezen van een levenspartner physisch en psychisch', *Afkomst en Toekomst* 3 (1937), 20–33.
7 K. Astel, 'Verwirklichung rassehygienischer Wohnungspolitik in Tübingen', *Afkomst en Toekomst* 4 (1938), 78–83; E. Rüdin, 'Die neuen rassenhygienischen Gesetze in Deutschland von medizinischen Standpunkt aus', ibid., 84–96.
8 J. Boeke, 'Iets over "rassen"', *Afkomst en Toekomst* 5 (1939), 193–202.
9 J. Ch. v. Schouwenburg, 'Overzicht van de ontwikkeling der eugenetica in Duitsland', *Afkomst en Toekomst* 5 (1939), 108–24. Hagedoorn: cf. Noordman 1989, 110.
10 W. F. H. Ströer, *Experimentelle Untersuchungen über die Mundentwicklung bei den Urodellen* (Berlin: Julius Springer, 1933); W. F. H. Ströer, 'Familiäres Auftreten von Reihen erblicher Hand- und Fuszabweichungen', *Der Erbarzt* 4 (1936), 22–6; W. F. H. Ströer, 'Ueber das Zusammentreffen von Hasenscharte mit ernsten Extremitätenmiszbildungen', *Der Erbarzt* 7 (1939), 101–4.
11 J. Mengele, 'Sippenuntersuchungen bei Lippen-Kiefer-Gaumenspalte', *Zeitschrift für menschliche Erblichkeits- und Konstitutionslehre* 23 (1939), 17–42.
12 Netherlands Institute for War Documentation (NIOD): press cuttings KB I 7743 (W. F. H. Ströer).
13 Rockefeller Archive Center, Sleepy Hollow, New York: Rockefeller Fellowship Cards.
14 NIOD: Nederlandsche Kultuurraad 111 E6, cv Ströer. For De Burlet cf. K. van Berkel, *Academische illusies. De Groningse universiteit in een tijd van crisis, bezetting en herstel, 1930–1950* (Amsterdam: Bert Bakker, 2005), 285–7.
15 National Archive, CABR 37770 (W.F.H. Ströer); NIOD: 77-206/207 (register Nederlandsche SS/Germaansche SS in Nederland).
16 N. C. K. A. In: 't Veld, *De SS en Nederland. Documenten uit SS-archieven*, vol. I ('s-Gravenhage: Martinus Nijhoff, 1976), 247.
17 W. F. H. Ströer, 'Vruchtbaarheid en onvruchtbaarheid', *Afkomst en Toekomst* 5 (1939), 146–52; M. J. Sirks and G. W. Kastein, 'Inleiding', M. J. Sirks and G. W. Kastein (eds) *Geneeskunde en erfelijkheid* (Lochem: De Tijdstroom, 1941), 9–13.
18 G. P. Frets, 'Geneeskundig onderzoek voor het huwelijk', *Afkomst en Toekomst* 7 (1941), 267–72; F. Schrijver, 'Beschouwingen over het vraagstuk van het geneeskundig onderzoek vóór het huwelijk', ibid., 273–85; H. J. T. Bijlmer, 'Het geneeskundig onderzoek vóór het huwelijk', ibid., 286–95; A. M. Benders, 'Opmerkingen in zake geneeskundig onderzoek vóór het huwelijk', ibid., 349–51.
19 L. de Jong, *Het Koninkrijk der Nederlanden in de Tweede Wereldoorlog*, vol. VI ('s-Gravenhage: Staatsuitgeverij, 1975), 592–3.
20 An., 'De erfelijkheid van normale en abnormale morphologische kenmerken', *Geneeskunde en erfelijkheid*, 75–100.

21 Cf. M. J. Sirks and P. J. Waardenburg, 'Introduction', M. J. Sirks and P. J. Waardenburg (eds), *Geneeskunde en erfelijkheid* (Lochem: De Tijdstroom, 1949), 8.
22 NIOD: 77-206/207; National Archive, CABR 76073 (J. A. van der Hoeven): interrogations by the Political Intelligence, 20 September 1946 and 12 March 1947.
23 National Archive, CABR 24043 (A.W. Wittermans).
24 Letter from O. Hofmann to H. Himmler, 30 September 1940, In: 't Veld 1976, 529.
25 NIOD: Nederlandsche Kultuurraad 111 E6: letter from Ströer to G.A.S. Snijder 27 July 1941.
26 NIOD: Nederlandsche Kultuurraad 111 E6: letter from Van der Hoeven to Snijder Wintermonth 1940.
27 L. Hajema, 'Ströer was nationaalsocialist uit overtuiging', *Universiteitskrant Groningen* 22 October 1987. Cf. also Van Berkel 2005, 360–1.
28 CABR 24043: letter from Van der Hoeven to Ströer, 22 May 1942; 76063: interrogation of Van der Hoeven. 20 September 1946.
29 CABR 37770.
30 CABR 24043: letter from Van der Hoeven to Ströer, 22 May 1942.
31 CABR 76073: letter from Van der Hoeven to De Lenselink 2 July 1942.
32 CABR 24043: letter from Van der Hoeven to Ströer 22 May 1942.
33 CABR 24043: letters from Van der Hoeven to Ströer 4 September and 15 November 1942.
34 NIOD: NSB 164d.
35 Bundesarchiv Berlin-Lichterfelde: NS/21 No. 136.
36 H.-W. Schmuhl, *Grenzüberschreitungen. Das Kaiser-Wilhelm-Institut für Anthropologie, menschliche Erblehre und Eugenik 1927–1945* (Göttingen: Wallstein Verlag, 2005), 359–60.
37 H. Schellekekens and R. P. W. Visser, *De genetische manipulatie* (Amsterdam: Meulenhoff, 1987), 93–4.
38 Schmuhl 2005, 481–2.
39 CABR 76073; 12142.
40 NIOD Doc I-1659 (Prof. dr. W. F. H. Ströer); Van Berkel 2005, 466–7.
41 Tribunaal Arrondissement Groningen sentence, 16 October 1947.
42 Bijzonder Gerechtshof Amsterdam sentence, 2 December 1947; CABR 24043: police statement 's-Gravenhage 4 November 1948.
43 A. W. Wittermans, 'De erfelijkheidsprognose van schizofrenie', *Nederlandsch Tijdschrift voor Geneeskunde* 87 (1943), 1217–22.
44 H. W. Scalongne, 'Rasverzorging', *Volksgezondheid* 2 (1942), 22–3.
45 P. J. Waardenburg, 'Een Nederlandsche wetenschappelijke belichting van rassenvraagstukken bij den mensch', *Mensch en Maatschappij* 16, (1941), 1–24.
46 P. J. Waardenburg, *Het rassenvraagstuk in onzen tijd. Een biologische toelichting* (Arnhem: Van Loghum Slaterus, 1937), 222.
47 G. W. Hylkema, *Ras en toekomst* (Amsterdam: De Amsterdamsche Keurkamer, 1942), 11–12.
48 J. H. R. van der Pas, 'De rasgebondenheid onzer wetenschap', *Volksgezondheid* 3 (1943), 61–3.
49 National Archive: CABR 73992 (J. C. Nachenius).
50 J. C. Nachenius, 'review of Hylkema, Ras en toekomst', *De Volksche Wacht* 6 (1942), 388–90.
51 Cf. Schmuhl 2005.
52 M. A. van Herwerden, 'Erfelijkheidsverschijnelen bij den mensch en eugenetica', *Nederlandsch Tijdschrift voor Geneeskunde* 68 (1924), 65–76; P. J. Waardenburg, *De biologische achtergrond van aanleg, milieu en opvoeding*, vol. I (Groningen: Noordhoff, 1927), 6.
53 W. F. H. Ströer, *Het ontstaan der rassen* (Amsterdam: Hamer, 1944), 16–17: W. F. H. Ströer, 'review of J. de Vries, Die geistige Welt der Germanen', *Rashygienische Medeelingen* 1 (1944), 106–7.

54 W. F. H. Ströer, 'De erfelijkheid van morphologische kenmerken', *Afkomst en Toekomst* 6 (1940), 101–8, especially 103.

55 NIOD: Departement van Volksvoorlichting en Kunsten 120e, Medisch Front guidelines.

8 The birth of a modern instrument and its development during World War II

Electron microscopy in Germany from the 1930s to 1945

Falk Müller

The electron microscope was on the verge of a very promising future when war broke out in 1939. The new technology seemed applicable in every field in which the visualisation of sub-light microscopic structures promised to provide answers to crucial questions. In some cases, vastly overrated expectations were raised. Some accounts celebrated electron microscopy as a tool that was about to knock at the door of atomic resolution and would, on a molecular basis, eventually fulfil the dream of early modern mechanists to visualise and make comprehensible the inner structures and mechanisms of organic and inorganic bodies alike.[1] The electrical engineers Ernst Ruska (1906–1988) and Bodo von Borries (1905–1956) had been engaged on the construction of an electromagnetic electron microscope since the early 1930s, first as students at the Technical University of Berlin and from 1937 onwards as employees of Siemens & Halske. The first type of their customised '*Übermikroskop*' or 'Super Microscope' was delivered to the IG Farben plant in Hoechst in late 1939.[2]

On 22 June 1938, Helmut Ruska (1908–1973) gave a talk to the Berlin Medical Society and showed the first images of viruses. His talk was published jointly with his brother Ernst and Bodo von Borries.[3] In a combined effort, the lecture and the publication eventually caused a public sensation that even reached as far as the political leaders of the Reich.[4] Shortly afterwards, IG Farben placed an initial order for three and later four instruments. However, Siemens' hopes of early commercial success with the electron microscope were dashed by the developing war and the appearance of competitors. In 1939, Manfred von Ardenne (1907–1997) had completed a high-resolution, multi-purpose electron microscope. While remaining an expensive laboratory appliance, this instrument was seen by contemporary observers as a sophisticated and innovative technical achievement. At the end of 1939, Hans Mahl (1909–1988), at the central laboratory of the *Allgemeine Elektrizitäts-Gesellschaft* (AEG) in Berlin, completed the construction of the first super microscope with an electrostatic lens system, an instrument capable of competing with the Siemens microscope if AEG had the chance to develop their

Figure 8.1 Picture of the first customised ‘Super Microscope’ that was delivered to the IG Farben plant in Frankfurt-Hoechst in 1939. In 1962 it was donated to the Deutsches Museum in Munich. (Picture courtesy of the Deutsches Museum.)

instrument into something more solid and reliable, to reach the status of development Siemens had already achieved.[5]

When representatives of Siemens praised the electron microscope in a letter to the Chief of the German Armed Forces Medical Services, Siegfried Handloser (1895–1954), in 1942 as an ‘important modern and universal research technology’,[6] they knew that this ‘modern technology’ still needed further development to unlock its potential. The construction of better instruments was just one of a series of tasks that had to be tackled. Of similar importance was the development of a set of preparation techniques allowing the visualisation of structures in objects that did not show sufficient contrast or that could not be investigated directly, such as the surfaces of solid bodies. With optional and probable applications but no clear proof of its usefulness, the electron microscope's career in the early 1940s shows that it was valued as an innovative and useful new research technology,[7] but not as a promising commercial endeavour.

The aim of this paper is to trace some strands of the implementation and further articulation of electron microscopy as a universally applicable research technology. Although not reaching its full formation until afterwards, this technology experienced astonishing development during the war; by 1945 around forty instruments had been produced and delivered to chemical factories, medical facilities and scientific laboratories in Germany and even abroad. Additional instruments were under construction or used in the research laboratories of Siemens & Halske, AEG and Manfred von Ardenne. Inevitably, the beginning of the war did influence the way electron microscopes were perceived and the circumstances under which research and production could proceed. By the end of the war, laboratories and production sites in particular had to be moved, researchers and technicians were drafted into military service and a general shortage of materials and labour force prevented a major expansion of effort. Before discussing some of the aspects of the development between 1939 and 1945, there will be a brief introduction to the foregoing developments at Siemens and AEG.

From Academic to Industrial Research: the Development of the Siemens 'Super Microscope'

It was no accident that electron microscopy matured in an industrial context. Only companies such as Siemens and AEG were able to supply facilities for the performance of the necessary research, for the coordination of the close cooperation of scientists from various backgrounds with engineers and technicians and for the construction and conjunction of various electrotechnical and mechanical components. In a report by the British Intelligence Objectives Sub-Committee (BIOS) on the development of electron microscopy in Germany during the war it was noted that 'the best work was done at the three institutes where workers of differing specialties combined in a team'. These conditions were only to be found at the Werner-Werke of the Siemens & Halske company, the private (but government sponsored) laboratory of Manfred von Ardenne and the laboratory of the AEG company.[8] Berlin served as the stage where neighbouring research activities resulted in a dense network of official and unofficial contacts, which were accompanied by a frequently unauthorised and unwelcome flow of information. In the presence of the emerging competition, Siemens and AEG both tried to hold or establish a good starting position for possible commercial exploitation. Owing to various strategies adopted by patent offices and the management, the competition, the quarrels and priority claims were partly compensated, repressed or ignored in the 1930s. In an unofficial agreement with AEG, for example, Siemens consented not to enter the field of electron optics, a field which in the early 1930s was dominated by the AEG research laboratory under its director Carl Ramsauer (1879–1955) and his deputy Ernst Brüche (1900–1985). While Siemens and AEG had cooperated not only in joint ventures such as Telefunken or Osram, but also in the establishment of cooperative research facilities, at the end of the 1930s the willingness to coordinate research activities ceased and both companies tried to retract resources and concentrate research in their own laboratories. In 1941, for example, Siemens sold its Telefunken shares to AEG and strengthened its own programme in radio engineering and wireless communication. At the same time, both companies stopped cooperating in several other fields.[9]

After Siemens had begun its successful venture into electron optics in 1937, AEG launched a polemical campaign against Siemens' attempt to harvest the fruits of a field that – in Ramsauer's and Brüche's eyes – had primarily been prepared by the effort of AEG researchers. When AEG suggested a concentration of the forces of the two companies as a solution to the growing conflict, Heinrich von Buol, chairman of the board of management of Siemens & Halske, answered somewhat ruggedly: 'The AEG – at least [its chairman of the board of management] Bücher – knows quite well why for years Siemens did not enter the field of electron optics. Such a propagandistic abuse of a noncommittal arrangement between Bücher and me does not encourage a repetition'.[10] Furthermore, Buol accused AEG of not having seized their chances when they presented themselves. The management had apparently failed to realise that even in 1936, as a 'typical representative' of electron optics, the electron microscope was 'a ripe fruit that only had to be picked'.[11] In contrast, Siemens managed to purchase and pool important patents.

The patent situation was difficult but indeed favourable for a management that showed a certain measure of recklessness and skill in launching and interpreting patents. Not only did Siemens hold rights of the former director of the scientific division of the Siemens-Schuckert-Werke, Reinhold Rüdenberg (1883–1961; Rüdenberg was forced to emigrate in 1935), but also those of Ruska and von Borries and some assigned by other researchers. Manfred von Ardenne, for example, transferred the utilisation of his patents and innovations to Siemens; some of these innovations were integrated into the further development of the Siemens microscopes.[12] Rüdenberg's patents were in many ways crucial since they laid claim to a very broad protection, not only of electromagnetic but also of electrostatic types of electron microscopes. While these patents were challenged by AEG and other companies, and although they were granted in Germany only after the war, they contributed to the provision of a protected space for the development of the Siemens microscopes.

In 1937, Siemens & Halske built up the Laboratory for Electron Optics in Berlin-Spandau with Bodo von Borries and Ernst Ruska as directors. In the years that followed, the company developed a logistical superstructure that allowed the further improvement of the electron microscopes and the implementation of electron optical methods in a variety of different scientific fields. In 1938 and 1939, two experimental super microscopes with a 60 kV acceleration voltage were installed in the new laboratory; in the same year another instrument with 120 kV was added. Ernst's brother, the physician Helmut Ruska, was invited to work out medical and biological applications for the new instrument. In order to promote its use in different scientific areas as quickly as possible, Ruska and von Borries suggested the establishment of a visiting laboratory for research work, the 'Laboratory for Super Microscopy', which eventually opened in 1940.[13] After Bodo von Borries had married Hedwig Ruska in 1937, the further development of electron microscopy at Siemens turned out to become a successful family enterprise.

The Laboratory for Super Microscopy was equipped with several fully developed electron microscopes, which were used by guest researchers and for educating new customers; in cooperation with both German and foreign scientists around two hundred papers were published by the end of 1944.[14] The new laboratory had several tasks to fulfil: it made instruments available to researchers who had no access to an electron microscope; it served as an advertising rostrum or 'shop window' for the further diffusion and public approval of the new technology; and it allowed Siemens to keep in close contact with its customers and to some extent even control their activities.[15] The Laboratory for Super Microscopy was led by Helmut Ruska, who served as an assistant to Richard Siebeck (1883–1965), the director of the I. Medical Clinic of the Charité, and by Gustav Kausche (1901–1960), from the Biologische Reichsanstalt für Land- und Forstwirtschaft.[16] As early as 1936, Siebeck had provided Ernst Ruska and von Borries with a positive assessment of the future application of electron microscopy in medical research; this expertise served as a decisive factor in Siemens' decision to support the further development of electron microscopy. Even after Siebeck left Berlin to become director of the Ludolf Krehl Clinic in Heidelberg, Helmut Ruska and Gustav Kausche's successor as co-director of the Laboratory for Super Microscopy, Carlheinrich Wolpers (1906–2003), maintained their position as Siebeck's assistants.

Siemens & Halske developed the most successful strategy for gathering expertise and interpreting and actively shaping the field of patents for its own purposes. This and the construction of a tight organisation that would allow instrument design, research on possible applications, the testing of new instruments and methods and the training of customers and new personnel to take place at one and the same site were the main reasons for Siemens' success, apart from the personal engagement of von Borries and Ernst and Helmut Ruska.

Electron Microscopy as a 'German Science'

From 1931 onwards, Ernst Brüche and his colleagues at AEG's central research laboratory constructed and used emission electron microscopes to visualise and magnify the surface structures of cathodes and hot pieces of metal. Owing to the laboratory's practical focus on the development of electrostatic devices, the construction of the electron microscope was beset with technical problems that could not be solved satisfactorily for several years.[17] In addition, the electron microscope was merely seen as one of several possible applications of electron optics and was not at the centre of interest. The early 1930s were, for example, dominated by the development of Braun tubes for the construction of oscillographs. As another reason for the delay in the construction of the first AEG super microscope – particularly put forward in their later dispute with Siemens – Ramsauer and Brüche mentioned their engagement on weapons research.[18] In 1934, Ramsauer decided to foster the construction of an electronic image intensifier rather than to continue the development of electron microscopes.[19] This decision was in line with the general policy of AEG, a company with a long tradition in military research: as essential components of night vision devices, the research on image intensifiers was promoted by the *Reichswehr*. Much of what happened at the research institute in the following years was in accordance with the German programme of rearmament and Göring's first Four-Year Plan.[20] As a result of the recovery of the German economy and the military engagement, the research institute's workforce increased from 140 in 1929 to 314 in 1936, to 600 in 1938 and finally 863 in 1944.[21] As director and leading scientist of the rapidly expanding central laboratory of one of the most important German companies, Ramsauer and

Brüche on the one hand experienced considerable freedom, while on the other they were confronted by growing responsibilities. Ramsauer in particular acted as a buffer and mediator between the physics community, party politics and the interest of the military and the state in Nazi Germany.[22]

Eventually, construction of the first AEG super microscope was not abandoned, but it was severely delayed. Nonetheless, when Hans Mahl of the AEG laboratory presented the first type of the electrostatic super microscope at the end of 1939, it was immediately seen by Siemens as an inconvenient competitor; this was particularly the case after the first instrument was donated to the Robert Koch Institute in Berlin and following the celebration of Mahl's achievement in the popular *Berliner Illustrierte Zeitung* under the lurid title: 'Deadly enemy of life tracked down by the new super microscope'.[23] For Siemens the situation was clear and non-negotiable: AEG should not be allowed to produce and sell its instruments.

Ernst Brüche had previously claimed to have been one of the co-inventors of electron microscopy. As (co-)author of several books in which electron optics was put on a firm footing,[24] he believed this new science had been fully developed only under his guidance. Now it seemed as if he were gradually losing control of the situation. On behalf of his own and his laboratory's reputation, he and his colleagues started a campaign in which they tried to highlight their own contributions and confronted the public with evidence aimed at exposing Siemens' claims on the priority and superiority of their achievements as weak and baseless.[25] Brüche and Ramsauer were driven by the belief that the production and operation of electrostatic electron microscopes were much easier and that they were cheaper than electromagnetic instruments and therefore more compatible with wartime conditions. Apart from other factors and in contrast to electromagnetic instruments, electrostatic instruments were not sensitive to voltage fluctuations, for example, and could be built in a more compact form.[26] There were, of course, many disadvantages as well, not mentioned by AEG.

AEG had some experience of the production of high vacua, but was somewhat less familiar with the high precision necessary for the production of the lenses and the construction design of mechanical devices such as the microscope's object stage. Since the object stage was to be moved and calibrated in the focusing process, electrostatic microscopes were very sensitive to a precise mechanical construction. In 1941, AEG ordered the fabrication of two object lenses at Carl Zeiss in Jena, a company with a long tradition in precision mechanics.[27] In May 1942, a contract between the two companies provided for a joint venture in which Carl Zeiss would take over the production and assembly of the lens system and the mechanical parts of the microscope, produce optical microscopes for the magnification of images shown on the screen and take responsibility for sales and distribution; both companies agreed to foster further research.[28]

AEG knew and Zeiss suspected that the patent situation was not resolved. Since a peaceful agreement with Siemens seemed to be out of reach, Brüche's hopes rested on the intervention of the head of the Reich Ministry for Armaments and Munitions, Albert Speer, and other government officials. In a conversation with members of Carl Zeiss staff, Brüche even claimed that the Reich Ministry of

Figure 8.2 Construction of an AEG super microscope. (Picture courtesy of the State Museum for Technology and Work, Mannheim.)

Economics, the Reich Ministry of Education and others had encouraged AEG to ignore the patent situation.[29] The researchers and the management at Siemens suspected that the driving force behind these government interventions was not a general interest in the further promotion of electron microscopy, but AEG management, which was using its friendly contact with many highly ranked politicians, military and party leaders for its own benefit.

As the negotiations with Carl Zeiss proceeded, Brüche and Ramsauer were engaged in a fierce struggle with members of the Siemens group involving various allies outside the laboratories and even the most senior management of the two companies. At the heart of the quarrel were several issues that cannot be described in detail here; most important was the question whether the initial patents of Siemens also covered the development and production of electrostatic electron microscopes, a field in which Siemens had never shown a noticeable interest. The quarrel spilled over into the public arena between 1940 and 1944 when both sides published long and polemical articles in newspapers and magazines. This public argument only stopped when many magazines ceased publication and once the government finally banned the publication of polemical articles in scientific magazines.[30]

In private communications with Ernst Ruska, Brüche appealed to what most German researchers normally emphasised when talking to government officials: that the electron microscope was primarily a German development, a 'German science', and that its advanced status should not be light-heartedly given away to other nations by excluding half of the active researchers. Brüche suggested close cooperation and a split between the production of electrostatic (AEG) and electromagnetic (Siemens) microscopes. Siemens' management responded to such claims by saying that the situation in Germany would not improve if the forces of competing companies were concentrated. As they saw it, the actual fight showed that the competition guaranteed a fertile climate that fostered the permanent production of ideas and the continual introduction of innovations.[31] AEG's production capacity was not needed since Siemens could handle the demand with their own resources; and while Siemens would give away their advantages, AEG had nothing substantial to offer. In internal notes, Siemens had to acknowledge that they were not able to satisfy the demand; as a possible solution, the management discussed the development and construction of simpler and smaller instruments, such as a compact microscope running with only 60 kV and an instrument using permanent magnets instead of electromagnetic ones.[32] In March 1943, in response to pressure exerted on Siemens by the High Command of the Army and the Army Ordnance Office, Siemens eventually offered AEG a licence for limited manufacture of electron microscopes. It was, however, clear that this exception was restricted to wartime and use within AEG research facilities.[33]

Making and Selling Electron Microscopes in Wartime

In spite of these problems, quarrels and the developing war, Siemens and AEG electron microscopes were soon to be found in several places in Germany and abroad. While AEG only managed to deliver very few super microscopes,[34] Siemens produced a total of thirty-eight instruments by the end of the war.[35] Siemens microscopes were installed in universities and research institutes and in plants of IG Farben Industries in Hoechst, Leverkusen, Bitterfeld and Wolfen, where they were used for materials science, such as the investigation of film emulsions, rubber, fibres or the properties of pigments.[36] Some instruments were delivered to Kaiser Wilhelm Institutes, such as the Institute for Silicate Research[37] and the Institute of Metal Research. The director of the Kaiser Wilhelm Institute of Physical Chemistry, Peter Adolf Thiessen, fostered the application of electron microscopy in the chemical sciences and purchased an electron microscope for his institute in 1942. Thiessen's own field of research was colloid chemistry, where he was used to more indirect methods (such as ultramicroscopy) of visualising sub-light microscopic particles. As head of the division of inorganic chemistry on the Reich Research Council he approved the acquisition of instruments for university institutes in Graz (Armin Dadieu), Stuttgart (Robert Fricke), Vienna (Ludwig Ebert) and elsewhere.[38] In 1944, an electron microscope was delivered to Strasbourg to a former student of Thiessen's, Dietrich Beischer, who held a chair of inorganic chemistry at the university. Beischer had been using electron microscopes in the investigation of colloids since the mid-1930s.[39] Further instruments were delivered to the universities in Tübingen, Freiburg and Heidelberg for medical, chemical and physical research. In 1943, one instrument was even sold and delivered to the Swedish biochemist and later Nobel prizewinner Arne Tiselius in Uppsala. The BIOS Report indicates that another instrument was found by allied forces in the *Istituto Superiore di Sanità* in Rome.[40]

The possibilities for enhancing production depended on the workforce available. Importantly in this respect Rudolf Mentzel had already, in November 1939, rated electron microscopy on behalf of the Reich Research Council as highly important to the war effort.[41] This status was approved several times up to 1945 and enabled the Siemens group to gradually increase its workforce. A request for foreign workers by Siemens, submitted in January 1943, was approved in order to enhance production. As additional employees Siemens requested: six mechanics, six lathe operators, four metalworkers, two cutters and four electricians.[42] An internal report for the year 1943/44 lists the following development of the workforce: from two clerks in 1937 (Ruska and von Borries) it increased to ten clerks and ten wage earners in 1937, forty-five and sixty-two in 1943 and forty-four and ninety-seven in September 1944; at that time four clerks and eleven wage earners were drafted into military service, while twenty-two of the wage earners were foreign workers.[43] On 1 April 1945 there were forty-seven clerks (five of whom were doing military service) and 110 wage earners (thirteen of whom were doing military service and fifteen foreign workers).[44] These numbers include the staff of the Laboratory for Super Microscopy and those workers who were employed in the production of Braun tubes and other electron optical devices. The important-to-the-war-effort status could not, however, prevent important researchers such as Helmut Ruska and Carlheinrich Wolpers being drafted into military service for several months.

In 1943, orders were received for thirty-nine instruments, but owing to a lack of workforce only ten could be produced and delivered. The company calculated a loss of about 3.5 million Reichsmark.[45] An internal note from November 1945 listed thirty-one instruments that were ordered but not produced during the war, some of them owing to missing or inappropriate urgency numbers. Orders came, for instance, from the Statens Seruminstitut Kopenhagen, the A.B. Scientia at Lund University, the Kliment Ohridski University in Sofia, Mitsubishi Shoji Kaisha in Tokyo and the Institute Pasteur in Paris.[46] The highest urgency levels were only granted in collaborations with powerful government or military agencies. In most cases, they were approved if the proposal was connected with the production and delivery of an electron microscope for a specific institution or person and not for further improvements of the instruments and methods; another decisive factor was whether the corresponding agencies were able to supply the necessary materials, such as special steels or rare non-ferrous metals.

Owing to agreements with other institutions, even the development and construction of new types of apparatus continued during the war.[47] In September 1944, the Siemens institute, for example, applied for funding for several research projects, in particular the construction of an instrument that was smaller and easier to handle and of an instrument with a higher voltage. These projects were approved by the Reich Research Council in October 1944. Eventually, the acceleration voltage of the Siemens microscope was increased several times during wartime.[48] In a project supported by the Air Force and the Reich Air Ministry, Ernst Ruska conducted research on a magnetostatic instrument and on a new type of electron microscope with the very high acceleration voltage of 750 kV. For these projects the Ministry provided a high urgency number and approved a sum of 24,000 Reichsmark for the following 12 months.[49] In an attempt to retain control of the project, von Borries suggested in a letter that this sum should be spent on materials only; the expenses for the development and construction would be covered by Siemens & Halske.[50] Such arrangements permitted the continuation of research projects and the allocation of rare materials on behalf of a superior and powerful agency while the company kept the initiative and the copyright.

Research in Times of War

Apart from these attempts to further improve the instrument's performance, its usability and reliability, the few available instruments were used intensively in collaborations between various institutions[51] and – in an act of interdisciplinary cross-fertilisation – between researchers from various disciplines such as biology, chemistry, mineralogy, physics or the medical sciences. An overview of the research being done at the AEG laboratory or with AEG instruments was given in the book *Zehn Jahre Elektronenmikroskopie*, which was republished in extended second (1942) and third (1943) editions; the 1943 edition was a richly illustrated book that even included stereoscopic images and specific stereoscopic glasses for their examination. In various sections it gave an account of advances in methodology and preparation technique and of research being done in metallurgy, metallography, the chemistry and physics of colloids and surfaces, zoology and various fields of medical research.[52] In a similar way, Manfred von Ardenne collaborated with researchers from various fields, such as the investigation of colloidal catalysts,[53] and he published broadly on the improvement of his own instruments and the introduction of new methodologies and devices.[54] As early as 1941, a research group was founded by the Stahl-Elektro-Union in cooperation with Siemens focusing on the investigation of steel with the super microscope. It was financially equipped with a fund of 25,000 Reichsmark.[55] At Siemens, von Borries in particular had been active in the development of electron optical methods for the investigation of surface structures and breaking edges of solid objects.[56] At the Laboratory for Super Microscopy specific instruments were reserved for the application of his methods. Later, two Siemens super microscopes were delivered to Krupp in Essen and to the Geisweider Eisenwerke near Siegen.[57]

In response to an attempt by Peter Adolf Thiessen to ensure a specific urgency number for the further development of electron microscopes, the Reich Ministry for Armament and Ammunition sent a questionnaire to Siemens. In its reply Siemens summarised those projects that were considered important to the war effort.[58] In the field of chemical research, the investigation of different kinds of soot for the production of synthetic rubber, the determination of appropriate particle sizes of pigments in colour production, the exploration of the structure and size of fibres in textile chemistry and the development of specific cement for fortifications were listed. In optics, the report mentioned the diminution and improvement of surface reflections of glasses applied in optical devices;[59] these experiments were of extraordinary importance for the armed forces, particularly since these investigations enabled the substitution of rare materials by more common ones. In surface technology, the report dealt with questions of superfinishing methods for wearless machine parts;[60] other projects touched on metallographic issues. In biological and medical research, the main focus was on virus research, the development of chemotherapeutic and disinfectant substances (the latter question in cooperation with Army Ordnance) and eventually on questions concerning the testing of warfare agents.[61] Another focus was on aerosols, on dust and smoke, which was particularly important for the improvement of gas mask filters.[62] Erhard Geißler mentions several institutions in which electron microscopes were used for research on measures against the possible use of biological weapons by Russian or Allied forces; in one case, however, it is not clear if experiments on the use of aerosols conducted by the meteorologist Günther Riedel at the Laboratory for Super Microscopy were connected to chemical and biological warfare.[63]

Before electron microscopes could also be used in medical and biological research, however, the practitioners had to develop specific preparation techniques and methods that allowed the visualisation of sub-light microscopic structures in organic objects. This required a willingness for basic research and the decision to release medical researchers from military service, which was not easily found in times of war. Instead, people argued that research of this kind should be postponed until the end of the war. Immediately after the beginning of the war Army administrators adopted a conservative or more hesitant attitude and preferred the application of light microscopes instead of introducing an untried new technology such as electron microscopy. Only in 1942 did the highest-ranked military physician, the Medical Inspector of the Army, Siegfried Handloser, cancel the decree of his predecessor Anton Waldmann that no research with electron microscopes should be conducted at the Military Medical Academy during wartime.[64]

During the months that Helmut Ruska and Wolpers were drafted into military service the research activities at the Siemens laboratory immediately slowed down. Amongst other reasons, their release coincided with a visit to the Laboratory for Super Microscopy in 1942 by the Swedish biochemist Tiselius. He and a colleague managed to identify and present the first images of the polio virus. The fact that visitors from abroad were conducting successful research while the leading German researchers were serving as physicians at the front was used as a decisive argument in attempts to release Ruska and Wolpers.[65] Incidentally, the whole situation also reminded Ruska and von Borries of earlier experiences, particularly of the years after they had left the Technical University in Berlin in 1933. While they were engaged on practical work in industry, other researchers had published results and were celebrated for achievements acquired with their apparatus. Ruska and von Borries were not keen on a repetition of such a situation and therefore focused much of their efforts on the further research and development of the research facilities under their control.[66]

In order to promote medical and biological research with the electron microscope, von Borries and Helmut and Ernst Ruska had previously, on 3 March 1941, turned to the State Secretary at the Ministry of the Interior, the Reich Health Leader, Leonardo Conti. In a letter they suggested the establishment of a Reich Institute for Structural Medicine in Frankfurt as a large-scale facility for electron microscopic research in the medical and biological sciences and as a necessary extension of the Siemens laboratories. Helmut Ruska and Wolpers were proposed as the institute's directors. In its first outline, this institute was supposed to house eighteen electron microscopes, over thirty researchers and some sixty other employees.[67] Conti, Frankfurt's mayor Friedrich Krebs and the head of the Nazi district of Hesse, Jakob Sprenger, were enthusiastic about these plans and offered generous support. The further development of these plans did not, however, proceed as desired, particularly after Frankfurt was hit by several air raids. A condensed version of such an institute was instead integrated into the Laboratory for Super Microscopy in Berlin. In close cooperation with the Military Medical Academy and other Army research institutions, Helmut Ruska and Wolpers conducted research on epidemic diseases such as spotted fever, typhus and other diseases that were threatening the soldiers on the Eastern Front.

Since the price of an electron microscope, around 70,000 Reichsmark for the 85 kV and 80,000 Reichsmark for the 100 kV instrument, was very high, efficient use was highly desirable; the full potential of these new instruments could only be exploited if research proceeded in close contact with the Siemens laboratory and if collaborations with additional scientists from various backgrounds could be arranged. Apart from the fact that the electron microscope manufacturers supported cooperative research as a means of advertising their instruments and accomplishments, they also gained valuable feedback on the performance of their instruments and on the demands of and recent developments in various fields of research. In the case of Siemens in particular it has to be kept in mind that the Ruskas and von Borries, although they enjoyed considerable freedom, ultimately had to comply with the company's requirements and commercial interests. In addition to research, members of staff also had to supervise the performance of instruments and practitioners at the new locations all over the Reich. Eventually, however, it is difficult to ascertain who actually used which instrument for what kind of research. After an assistant or operator had been trained at the Laboratory for Super Microscopy for some weeks or months, an institution could in principle operate without the Siemens staff knowing exactly what kind of research was being conducted.

In short, although various applications did come to light, the difficulties of controlling and judging the instrument's performance, the difficulties of interpreting the electromicrographs and the problem of comparing and adjusting the results to data found using other methods showed that electron microscopy was far from being a widely available and reliable method.[68] Asked whether he supported the extension of the production of electron microscopes and whether he thought the electron microscope was an instrument 'decisive for the outcome of the war', Siemens' chairman of the board of management, Heinrich von Buol, pointed out that only the right mixture and quantity of various devices and not the production of a single device could play such a role. In the case of the electron microscope von Buol decided in favour of further development and production. As a justification for his decision he suggested that as a research technology the super microscope made an abrupt surge in progress not only possible but highly probable: 'At the current stage of development we are about to cross the border into a new territory'.[69] It was the expectation of this step and the surprises that accompanied the entering of the new territory that made electron microscopy such a valuable tool: on the one hand, for giving the enemy a surprise; on the other, for protection from surprises given by the enemy.

Public Image and Propaganda

Apart from the endeavours to win the sympathy of the scientific and medical communities and the authorities involved, AEG and Siemens started a public campaign to promote the electron microscope. The electron microscope was presented as an achievement that contributed to the prestige of German science and was used in the war of propaganda against the USA in particular. Siemens' promotional campaign was further supported by the public sensation the instruments caused when the first images of viruses were published, and these results even reached the leading figures of the Reich. In early 1939, at a time when customised instruments were not yet available, the Major General Reich Commissioner for Health and Sanitation and personal physician to Hitler, Karl Brandt, discussed the possible ordering of three instruments on Hitler's behalf with the staff of the Siemens laboratory.[70] While Brandt continued to act as one of the principal supporters of the electron microscope in the Nazi hierarchy, Hitler remained an interested observer of its further development; at one point he would even comment on the pre-war discussions at Siemens whether and under what conditions electron microscopes should be exported to the USA. While Brandt preferred to restrict access to German scientists and to exploit the new technology's potential to the sole advantage of German science, he reported that Hitler and Joseph Goebbels had shown no general aversion to the export of these instruments; both, however, criticised the price of 70,000 Reichsmark, the price that was proposed for the German market, as being far too low.[71] In the following years many government and party officials, such as Albert Speer, Werner Osenberg, Rudolf Mentzel or Hitler's personal physician Theodor Morell,[72] were invited to inspect the laboratories.

In 1941, Max Knoll, Ernst Ruska, von Borries, von Ardenne, Mahl, Hans Boersch and Brüche were awarded the Leibniz Prize of the Prussian Academy of Science for the development of the electron microscope. In April 1943, von Borries and Ernst and Helmut Ruska were guests on their own TV show 'Das Übermikroskop'.[73] In 1943, Mahl received a donation of 5,000 Reichsmark for the development of the electrostatic super microscope.[74] In April 1944, Ernst Ruska had been awarded the Fritz Todt Prize in gold as one of the first laureates. The prize was donated by Hitler and awarded for inventions with an exceedingly high value for the development of new weapons and for other war efforts.[75] This award can be seen as a decoration of Ruska-the-engineer, representing the popular figure of the inventor as the leader and the man of the decisive deed at his best. In addition, it shows that the electron microscope as a 'modern technology' and its inventors and patrons had come very close to the very centre of power in Nazi Germany.

The electron microscope yielded Siemens more prestige than profit. Indeed, as von Borries suggested after the war, Siemens supported the development and production of electron microscopes primarily as a means of promoting the company's public image and to demonstrate its 'acceptance of cultural responsibility'.[76] In his study of the early development of electron microscopy in northern America, Nicolas Rasmussen suggests that the Radio Corporation of America (RCA) similarly took up the production of electron microscopes during the war less for commercial reasons[77] than for the prestige associated with such a technology. The management hoped that the prestige could be transferred to the firm's marginally successful television tubes.[78] Moreover, in the United States as in Germany the electron microscope was padded with martial terms during the war – 'powerful weapon of science', 'Shooting Electrons Shatter Barriers to Science March'; it was propagated as a symbol of the superiority of German science that had to be overcome[79] and it was decorated with suggestive examples such as the application of super microscopes in 'making cement for the Siegfried line'.[80]As far as they became known, publications of American authors were regularly used by German researchers to exert pressure on the politicians and science organisers.

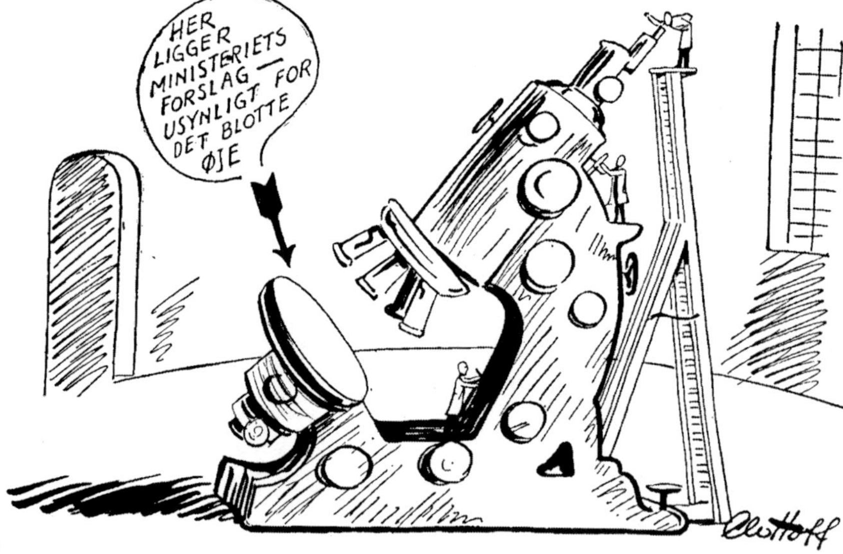
I Verdens stærkeste Mikroskop

I Tyskland er der konstrueret et Mikroskop, der forstørrer 30,000 Gange.

Formentlig vil det da endelig være muligt at øjne Ministeriet Staunings Forslag til Afhjælpning af Arbejdsløsheden.

Figure 8.3 There were, however, other ways of putting pressure on one's government. On 23 July 1938, commenting on the first public announcement of the Siemens super microscope, the Copenhagen newspaper *Berlingske Tidende* published a caricature presenting the new microscope as a solution to the problem of visualising Thorvald Stauning's (the Danish Prime Minister's) inadequate proposal for the lowering of the unemployment rate.

The members of both groups at Siemens and AEG were involved in advertising the new research technology and their own instruments abroad. In February 1940, Helmut Ruska, for example, gave a talk to the Society for Biophysics in Delft in The Netherlands.[81] In May 1943, von Borries lectured at the Istituto Superiore di Sanità in Rome.[82] Ernst Brüche gave four talks in Budapest, Hungary, in 1944.[83] In January 1944, Helmut Ruska travelled through France for about a week. His aim was to establish or revive contact with French scientists and physicians and to 'conduce to the French–German understanding'.[84] On his tour, which was organised by the German Institute in Paris, Ruska visited institutes and gave talks in Paris, Toulouse, Marseilles and Bordeaux. As he proudly reported, more than 900 listeners attended his talk in Paris; in the other cities he lectured in front of an audience of 200 to 250 people. His attempts to revive contact with French scientists seem to have been successful; Professor Constantin Levaditi and Dr Bonét-Maury at the Institut Alfred Fournier, Laboratoire de Syphilis et Chimiothérapie expérimentale, and Dr Nicolas Boulgakov at the Institut d'Hérelle, Laboratoire du Bactériophage, agreed to make their strains of virus and bacteria available to the Berlin group; Professor Levaditi even consented to providing preparations of penicillin.[85] And Ruska negotiated the delivery of an electron microscope with the president of the Institut Pasteur, Professor Tréfouël. An instrument had been ordered two years previously, but the responsible German agencies had refused to grant the necessary urgency number. As they saw it, 'cultural propaganda' would be the only reason for the delivery of such an instrument, while the contribution to the war effort was more than doubtful.[86]

The End of the War

On 24 and 25 November 1944, a workshop on the 'status and capacity of electron microscopy' took place in Berlin-Babelsberg. As mentioned in the invitation, every participant had to provide for their own accommodation and bring food ration cards for the meals.[87] The workshop was organised by the Surgical Adviser of the Army and Chief of the Office for Medical Science and Research, Paul Rostock, in cooperation with the Siemens group; its main focus was on medical research. Wolpers, Ernst Ruska and von Borries suggested a list of participants and appropriate topics. As they emphasised, they wanted to avoid a continuation of the debate about 'historical' issues and priority claims with the AEG research group. Therefore, they planned to exclude discussions of purely technical questions and instead focus on advances in preparation methodology, on actual progress in the application of electron microscopes in medicine and biology as well as its limits, or the generation of the electron microscopic image. On their list they dismissed Manfred von Ardenne owing to the sole focus of his talks on his own instrument; Ramsauer and Brüche were not included since they did not publish in the field of electron microscopy; Hans Mahl and a medical researcher who cooperated with the AEG laboratory were the only members of the AEG group to be put on the list. Still, as Ruska and von Borries underlined, their judgement was far from unbalanced since: 'it cannot be denied that by far the major share of all researches and applications of electron microscopy in the biological and medical field was conducted in the Siemens laboratory'.[88] Eventually, the participants included Ernst and Helmut Ruska, Bodo von Borries, H. O. Müller, Peter Adolf Thiessen, Hans Mahl, several medical researchers and, thanks to Rostock's intervention, Manfred von Ardenne.[89] The workshop was a last attempt to further promote electron microscopy in Germany, particularly in the medical sciences. A long report in the *Völkischer Beobachter* on the general development of electron microscopy and the specific problems discussed at the workshop shows the high public interest and the high credit party and government officials gave to this new technology.[90]

The workshop also showed that AEG's research and achievements were not sufficiently appreciated – that was at least the impression of Ramsauer and Brüche and one of the reasons why Ramsauer cancelled his participation. It was clear that Siemens had taken the lead, at least in the realm of medical and biological research, a field in which Brüche had already abandoned hope of being able to compete; his hopes rested on AEG successes in the material sciences. Here, a new method developed by Hans Mahl, the use of thin imprints or replicas of surface structures in order to make the investigation of solid objects possible, proved to be a major breakthrough.[91]

In December 1944, an internal report at AEG 'On the status of the development of electron microscopy' urged a decision in favour of a more focused engagement in electron microscopy and warned that the field would be completely lost to Siemens sooner rather than later:

> Siemens has gained a considerable advantage in recent years and, although the patent situation has improved, it does not seem likely we can catch up at a later date. The medical practitioners at least will have become accustomed to Siemens, as has been made alarmingly clear at the workshop.[92]

Of course, people at AEG and Zeiss were not pleased by these developments. In January 1945, Ernst Brüche summed up the situation in a letter to the chief physicist at Carl Zeiss, Georg Joos:

> We will experience the same as the man who returned home from war only to notice that somebody else has taken his place, somebody who would not leave voluntarily. ... Siemens has produced and installed about 30 instruments, has mastered the teething problems, has smoothed out the weak points in the production process, and managed to accustom users to the electromagnetic type.[93]

In a letter to Jonathan Zenneck, the director of the Deutsches Museum in Munich, Ramsauer referred in a similar vein to AEG's military engagement; seemingly, Ramsauer showed symptoms of an incipient hangover:

> If today others take advantage of our inability to proceed in civil sectors such as the construction of the super microscope because we were completely absorbed by the rearmament programme, it seems to me as if a combatant would return home after the war and you would let him know: 'It's highly deserving that you served as a soldier, but your civil position has already been occupied by someone else'.[94]

In a last attempt to catch up, Brüche reported to Joos in January 1945 that AEG management had finally decided to take up the production of electron microscopes comparable to Siemens' engagement and that they had already chosen a plant where 20–25 instruments per year should be produced. As a contribution to the war effort, AEG's initiative came much too late; it can therefore only be interpreted as a preparation for the post-war era in which electron microscopy as a 'civil' technology was much more promising than other wartime engagements. It is not, however, clear how Brüche could still believe that the capacities for such an endeavour would be available. The correspondence between AEG and Zeiss was full of bad news on bottlenecks and delayed delivery of parts. An order for three objectives and projection lenses issued by Brüche in January 1944 was eventually delivered at the end of December; owing to more urgent orders from the Army, as Zeiss apologised. Quality inspection at AEG found fault with the partly rusty, partly distorted and largely incomplete case. Although the electrodes, on the other hand, were in perfect condition, it is clear that production no longer met Zeiss' highest requirements but took place under increasingly pragmatic conditions.[95] Other occupations at least prevented Georg Joos from an overenthusiastic engagement. In 1945, he reported to Brüche that he was very busy since, apart from other duties, he served as a company leader of the *Volkssturm*.[96]

In February 1945, the Laboratory of Electron Optics at Siemens received the last directions from the Reich Research Council to continue its work since the instruments were desperately needed.[97] At that time, parts of the laboratories and the main production sites had already been destroyed. Like their instruments, the Berlin community of electron microscopists had been scattered all over Germany. While Manfred von Ardenne and Ernst Ruska stayed in Berlin, von Borries left the city and rescued two microscopes, taking them to a village near Herford in western Germany; these instruments and some additional equipment served as the basis for the new beginning of electron microscopic research at Siemens. The Laboratory for Super Microscopy was destroyed in October 1944. One of the electron microscopes was transferred to the military research facility at the Sachsenburg in Saxony in late 1944;[98] Helmut Ruska moved other parts of its equipment to the island of Riems on the Baltic coast, where he continued conducting virus research. Ernst Brüche and parts of the AEG research laboratory had moved several times before they finally settled in the city of Mosbach near Heidelberg, waiting for a new beginning.

Discussion

In their BIOS Report, 'Electron Microscopy in Germany', British Intelligence acknowledged the very high 'standard of technique and knowledge of the best of the German workers' but they do not seem to have been particularly impressed by the overall developments when they concluded: 'The use of the electron microscope yielded no new fundamental knowledge of importance'.[99] While other fields of science and technology that were developed during the war have attracted a great deal of attention, electron microscopy seems to have been a comparably 'neutral' and far less exciting field for historical investigations. It did not result in cataclysmic developments, and many discoveries and innovations were simultaneously made in other countries.

Nevertheless, the researchers at both companies were proud of their achievements and they definitely felt that the electron microscope could serve as a figurehead of German science and technology. The researchers also knew that this technology and its applications in many cases still belonged to the realm of basic research. While some electron optical systems reached an advanced status, others were in a nascent phase; while the applications of electron microscopy in new fields of study aroused the interest of a growing number of researchers, the provision of preparation techniques, the interpretation of the images produced and the estimation of their value for technological applications and the production process were still very difficult.

Siemens and to a much lesser extent AEG were able to continue the construction and production of electron microscopes until the end of the war. While AEG was left behind owing to other commitments and the hesitant attitude of its management, Siemens successfully applied various strategies that helped to compensate for inconveniences induced by the war and allowed the researchers considerable freedom. Consequently, the Siemens group managed to establish electron microscopy as a modern and important research technology in various fields of science. One successful strategy was the establishment of new facilities such as the Laboratory for Super Microscopy, which served as a site for testing and promoting the new technology. Another important factor was the organisation of a tight communication between the practitioners and the core team around Ernst Ruska and von Borries; the 'family business' at Siemens, the role allocation and the division of labour between Bodo von Borries and Ernst and Helmut Ruska were decisive factors as well. The dominance of the Siemens group should not, however, be overemphasised. In conclusion, the competition between the two research groups (or three if von Ardenne is included) produced a rich material culture in which a variety of innovative systems and instruments were constructed and tested and which served as a strong basis for the development of electron microscopy in Germany after the war.

Acknowledgements

Much of this paper was written during the author's participation in the Scholar-in-Residence programme at the Deutsches Museum. The author also has to thank Florian Schmaltz for valuable information and critical remarks.

Notes

1 See e.g. 'Molecules, the inner constitution of germs and even different kinds of viruses can be visualised' ('Einbruch in das Reich des Unsichtbaren. Das Übermikroskop offenbart Geheimnisse', *Deutsche Allgemeine Zeitung*, 21 July 1938); 'Die Welt der Atome sichtbar!' *Der Neue Tag*, Prague, 20 March 1940; Manfred v. Ardenne, *Elektronen-Übermikroskopie* (Berlin: Springer, 1940), ix–x.

2 An electron microscope was called a 'super microscope' if it surpassed the limit of possible magnification and resolution of light microscopes. For a general introduction to the history of electron microscopy according to Ernst Ruska, see E. Ruska, 'The Early Development of Electron Lenses and Electron Microscopy', *Microscopica Acta*, suppl. 5 (Stuttgart: 1979); here another important pioneer is mentioned who does not appear in this paper, Max Knoll. For biographical notes on Ruska, see L. Lambert and T. Mulvey, 'Ernst Ruska (1906–1988), Designer Extraordinaire of the Electron Microscope: A Memoir', *Advances in Imaging and Electron Physics* 95 (1996), 3–62.

3 B. von Borries, E. and H. Ruska, 'Bakterien und Virus in übermikroskopischer Aufnahme', *Klinische Wochenschrift* 17 (1938), 921–5.

4 See e.g. reports in *Berliner Zeitung am Mittag*, 23 June 1938: 'Triumph deutscher Forschung. Übermikroskopie vergrößert 20000 fach. Dinge von denen man bisher nichts ahnte'; *Der Angriff* (Goebbels' newspaper in Berlin), 23 June 1938: 'Viren – 20000 fach vergrößert. Deutsche Physiker erfanden das Übermikroskop'; *Rheinisch-Westfälische Zeitung*, 24 June 1938: 'Vergrößerung – 22000 fach! Ein sensationeller Vortrag in der Berliner medizinischen Gesellschaft'; *Neue Freie Presse*, Wien, 26 June 1938: 'Erschließung des Unsichtbaren. Neue Welten im Elektronenmikroskop'.

5 In electron microscopes, specifically designed electric or magnetic fields are used as lenses which affect electron beams in a similar way as glass lenses affect beams of light. In the construction of electron microscopes, electrostatic and electromagnetic systems are alternative and competing modes of realisation. While there are many alternative systems which use electrons or electron beams for the magnification of material structures, here the focus will be on the development and application of transmission electron microscopy.

6 Copy of a letter from Flir (proxy at the head office of the Siemens Werner-Werke) and Bodo v. Borries to Handloser, 29 September 1942; MPG Archive, III, ZA 137 /18.

7 Although I referred to the electron microscope as a 'modern instrument' in the title, in the long run it might be more appropriate to use Terry Shinn's term 'research technology' for an instrument that developed its dynamic not in a single discipline but between several disciplines and as a device that absorbed and coordinated the activities of technicians and scientists in university and industrial contexts alike (T. Shinn, 'The Bellevue grand électroaimant, 1900–1940: Birth of a research-technology community', *Historical Studies in the Physical Sciences* 24 (1993), 157–87; for spectroscopy as a case study, see K. Hentschel, 'Spectroscopy or Spectroscopies?' *Nuncius* 27 (2002), 588–614).

8 Von Ardenne was praised for his 'outstanding in practical use of the instrument', Ernst Ruska for the 'theoretical design of [the] electro-magnetic instrument' and his brother Helmut for his 'biological, especially virus research'. The only person from the AEG institute mentioned was Hans Mahl and his metallographic research ('Electron Micro-

scopy in Germany', British Intelligence Objectives Sub-Committee, Final Report No. 1671, 10).

9 For an introduction to and a comparison of the style and policy of industrial research at Siemens and AEG see P. Erker, 'The Choice between Competition and Cooperation: Research and Development in the Electrical Industry in Germany and The Netherlands, 1920–1936', F. Caron, P. Erker and W. Fischer (eds), *Innovations in the European Economy between the wars* (Berlin/New York: de Gruyter, 1995), 231–53; 246. On the cooperation, the joint research activities and the exchange of patents in the field of light bulbs, radio and electron tubes, see G. Luxbacher, *Massenproduktion im globalen Kartell. Glühlampen, Radioröhren und die Rationalisierung der Elektroindustrie bis 1945* (Berlin: GNT, 2003).

10 Draft of a letter from Heinrich von Buol to Carl Ramsauer, undated (probably April 1943), MPG Archive, III, ZA 137 / 92.

11 Ibid.

12 G. Barkleit, *Manfred von Ardenne. Selbstverwirklichung im Jahrhundert der Diktaturen* (Berlin: Duncker und Humblot, 2006), 54–62.

13 The Laboratory for Super Microscopy opened at a ceremony in the presence of many government officials on 18 April 1940 (s. 'Das Übermikroskop als Forschungsmittel. Vorträge, gehalten anläßlich der Eröffnung des Laboratoriums für Übermikroskopie der Siemens & Halske A.G., Berlin', issued by Siemens & Halske (Berlin: de Gruyter, 1941)). It was later equipped with four super microscopes (C. Wolpers, 'Electron Microscopy in Berlin 1928–1945', *Advances in Electronics and Electron Physics* 81(1991), 211–9: 224, 216).

14 E. Ruska, 'Autobiography', T. Frängsmyr and G. Ekspong (eds), *Nobel Lectures, Including Presentation, Speeches and Laureates' Biographies, Physics 1981–1990* (Singapore: World Scientific, 1993), 354; *Veröffentlichungen aus den Laboratorien für Elektronenoptik und Übermikroskopie der Siemens & Halske AEG, 1938–1949*, (Berlin: Siemens & Halske, undated).

15 At this point, it would be interesting to compare Siemens' attempts to achieve hegemony in the field of electron microscopy with Nicolas Rasmussen's account of the 'picture control' as practised by the scientists at RCA (N. Rasmussen, *Picture Control. The Electron Microscope and the Transformation of Biology in America, 1940–1960* (Stanford: Stanford University Press, 1997)); this comparison will be made in a subsequent publication.

16 Kausche and his colleague Edgar Pfankuch had successfully purified viruses earlier in the 1930s which served as a decisive preliminary work for the visualisation of viruses; see G. A. Kausche, E. Pfankuch and H. Ruska, 'Die Sichtbarmachung von pflanzlichem Virus im Übermikroskop', *Die Naturwissenschaften* 27 (1939), 292–9).

17 Seemingly, these problems have been characteristic features in the development of electrostatic electron microscopes since they arose at other companies as well. For a social constructivist account of these problems in the American case, see G. C. Kunkel, 'Technology in the Seamless Web: "Success" and "Failure" in the History of the Electron Microscope', *Technology and Culture* 36 (1995), 80–103.

18 'I want to stress that in the years after the [political] change our electron optical laboratory was almost entirely absorbed by armament and war efforts while you were engaged in the further promotion of electron microscopy'. Brüche to Ernst Ruska, 15 May 1942, Archive of the State Museum for Technology and Work, Mannheim (LTA), Brüche II / 279.

19 Archive of Deutsches Technikmuseum, Berlin, AEG / 01951; B. Weiss, 'Rüstungsforschung am Forschungsinstitut der AEG bis 1945', H. Maier (ed.), *Rüstungsforschung im Nationalsozialismus. Organisation, Mobilisierung und Entgrenzung der Technikwissenschaften* (Göttingen: Wallstein, 2002), 109–41: 127.

20 See ibid. for a detailed account of the military research at the AEG *Forschungsinstitut* in the 1930s and 1940s.

21 Weiss 2002, 123; the scientific and technical staff increased from about 30 in 1928 to 121 in 1940. These numbers can be compared to the development at the central laboratory of Siemens where the staff increased from 269 in 1933 to 883 in 1944 (P. Erker, *Industrie-Eliten in der NS-Zeit* (Passau: Rothe, 1993), 52).

22 From 1941 to 1945 Ramsauer was the president of the German Physical Society. For information on the development of the DPG and Ramsauer and Brüche's role during the Nazi era, see D. Hoffmann, 'Between Autonomy and Accommodation: The German Physical Society and the Third Reich', *Physics in Perspective* 7 (2005), 293–329 and D. Hoffmann and M. Walker, *Physiker zwischen Autonomien und Anpassung. Die Deutsche Physikalische Gesellschaft im Dritten Reich* (Weinheim: Wiley-VCH, 2006).

23 'Todfeinde des Lebens vom neuesten Übermikroskop gestellt', *Berliner Illustrierte Zeitung* 51 (1939), 1907. The microscope has a long tradition as a symbol of the fight against or the visualisation of invisible enemies (C. Gradmann, 'Invisible Enemies: Bacteriology and the Language of Politics in Imperial Germany' *Science in Context* 13 (2000), 9–30); the electron microscope blended in perfectly with this tradition.

24 E. Brüche and O. Scherzer, *Geometrische Elektronenoptik. Grundlagen und Anwendungen* (Berlin: Springer, 1934); H. Busch and E. Brüche (eds), *Beiträge zur Elektronenoptik* (Leipzig: Barth, 1937); E. Brüche and A. Recknagel, *Elektronengeräte. Prinzipien und Systematik* (Berlin: Springer 1941); C. Ramsauer (ed.), *Das freie Elektron in Physik und Technik* (Berlin: Springer 1940).

25 Still valuable as an introduction to the German developments and the priority dispute between AEG and Siemens is L. Qing, *Zur Frühgeschichte des Elektronenmikroskops* (Stuttgart: GNT, 1995).

26 C. Ramsauer, 'Entwicklung der Übermikroskopie', *Jahrbuch der AEG-Forschung* 7 (1940), 1. Similar arguments were employed in a competition between the two rival systems taking place in the United States in the 1940s and 1950s. There, General Electric and RCA took over the roles of AEG and Siemens (G. C. Kunkel 1995). In the case of GE the close cooperation with AEG in the pre-war period (including the exchange of patents) might have been an important factor for such a specific focus. In March 1941 von Borries reported to the Reich Research Council that Siemens and RCA carried on direct negotiations (v. Borries to Reich Research Council, 26 March 1941, Bundesarchiv (BA)-Berlin, R 26, III, "Reichsforschungsrat" / 199).

27 Letter to Carl Zeiss, 28 June 1941, Carl Zeiss Archive, BACZ 23423. Another example is the microscope's lenses and apertures, which were produced at Zeiss with a tolerance of 0.01 mm (Carl Zeiss to AEG, 4 December 1942, Carl Zeiss Archive, BACZ 23423).

28 MPG Archive, III, ZA 137 / 76; Carl Zeiss Archive, BACZ 23423, report on a visit to the AEG laboratory, 4 December 1941. The cooperation between AEG and Carl-Zeiss (now in Oberkochen, West Germany) was renewed after the war and lasted until the early 1950s when AEG gradually withdrew from the development and production of electron microscopes.

29 He could not, however, provide any written evidence and the reporter was not sure if such orders really existed (Carl Zeiss Archive, BACZ 23423, report on a visit to the AEG laboratory, 9 November 1942). Not much later, in 1944, several government and party organisations tried to abolish the maintenance and secrecy of patents in order to enhance the war effort (K.-H. Ludwig, *Technik und Ingenieure im Dritten Reich* (Düsseldorf: Droste, 1979), 270 and 506).

30 In November 1944, the editor of the *Physikalische Zeitschrift* asserted as the reason why he had to decline the publication of an article by Ramsauer: 'Following a decree of the Reich Ministry of Education, articles of a polemic character or such dealing with priority questions should no longer be printed during the wartime', LTA, Brüche II / 279.

31 '[G]erade der jetzt bestehende Kampf führt ständig zu neuen Anregungen und Entwicklungen, während die Konzentration der Kräfte, die die AEG hervorhebt, praktisch nur dahin führen würde, daß noch jede Entwicklungsstelle wie bisher weiter entwickelt. Das Fehlen eines wirklichen Konkurrenzkampfes innerhalb der deutschen Industrie könnte hier eines Tages eher fortschrittshemmend werden', (Flir, to Wiegand, Siemens Patent Office, 11 March 1943, MPG Archive, III, ZA 137 / box 167).

32 Ibid.

33 Siemens & Halske to AEG, 23 March 1943, MPG Archive, III, ZA 137 / 38.

34 AEG delivered five instruments. Apart from the Robert-Koch-Institute, additional instruments were delivered to the KWIs for Silicate Research (Eitel), Biophysics (Rajewski) and Physical Chemistry (Thiessen). Similar to the situation at Siemens, AEG had to decline orders such as the request of the William G. Kerckhoff-Foundation, which wanted to donate an instrument to the research facilities of the German Air Force (letter from Brüche to Carl Zeiss, 7 September 1943; Carl Zeiss Archive, BACZ 23423). A list mentions fifty-four seriously interested institutes, government agencies and companies in Germany, in allied and occupied countries (undated, LTA, Brüche II / 279).

35 Wolpers 1991, 224.

36 A report on the use of the Übermikroskop at the IG Farben plant in Hoechst can be found in: F. Schmieder, 'Beispiel für die Anwendung des Übermikroskops auf chemisch-technische Fragen', *Das Übermikroskop als Forschungsmittel* (Berlin: De Gruyter, 1941), 67–77.

37 One focus of the institute's research activities important to the war effort was the investigation of concrete, which was used in the construction of motorways and fortifications (W. Eitel, 'Das Übermikroskop als Instrument für die quantitative Messung in der Silikatforschung', *Das Übermikroskop als Forschungsmittel* [Berlin: De Gruyter, 1941], 48–66).

38 U. Deichmann, *Flüchten, Mitmachen, Vergessen. Chemiker und Biochemiker in der NS-Zeit* (Weinheim et al.: Wiley-VCH, 2001), 234. As one of the earliest and most serious supporters of electron microscopy in Germany, Thiessen arranged grants for Manfred von Ardenne as well (see BA-Koblenz, R 73 / 10090, especially a long affirmative report on Ardenne's researches from 7 July 1939).

39 Beischer's application for an electron microscope was already approved in November 1941 for investigating the 'morphology of protective colloids' (Rudolf Mentzel, Reich Research Council, to Beischer, 6 November 1941, BA-Koblenz, R 73 / 10223) but it was only delivered in 1944.

40 BIOS 1671, 6.

41 Letter from Rudolf Mentzel as representative of the president of the Reich Research Council to Ernst Ruska, 7 November 1939: 'Hereby I confirm that the electron microscope you developed (*Übermikroskop*) is a tool for the investigation of numerous war related issues. I emphatically support the further development of this valuable instrument, which is exceedingly important to the war effort', (MPG Archive, III, ZA 137 / 18).

42 Flir und v.Borries to the head of the 'division of communications technology and measuring instruments', Gustav Leifer, 26 July 1943, MPG Archive, III, ZA 137 / 42. It was emphasised that these workers should be highly qualified; it is not noted where they came from, on what terms and under what conditions they actually worked. For a general discussion of foreign and forced labourers at Siemens, see W. Feldenkirchen, *Siemens, 1918–1945* (München: Pieper, 1995), 203–12. On Leifer's role see ibid, 214.

43 Internal report on the institute's development in 1943/44, MPG Archive, III, ZA 137 / 42.

44 Internal report on the institute's development in 1944/45, MPG Archive, III, ZA 137 / 42. It is, however, not clear if employees paid by the Reich Research Council or other

agencies are included. Most probably the number of researchers and technical assistants was higher.

45 Flir und von Borries to the head of the 'division of communications technology and measuring instruments', Leifer, 7 July 1943, MPG Archive, III, ZA 137 / 42.

46 'Liste der noch offenen Bestellungen', Berlin, 7 November 1945; MPG Archive, III, ZA 137 / 38.

47 Some of the technical improvements of the Siemens microscope are summarised in A. W. Agar, 'European Commercial Electron Microscopes', T. Mulvey (ed.), *The Growth of Electron Microscopy*, vol. 96, *Advances in Imaging and Electron Physic* (San Diego et al.: Academic Press, 1996), 415–584: 417–24, 427.

48 'If Siemens had adopted the numbering convention which was adopted by all companies after the war, there should have been three distinct models launched in this period, to distinguish the changed specification'. (Agar 1996), 428.

49 Flir to Buol, 18 September 1942, Siemens Archive (SAA), 11 Flir / Lg 29.

50 von Borries to the research authorities of the Reich Air Ministry, 5 August 1942, SAA 11 Flir / Lg 29.

51 This is true even for the instruments mounted in industrial laboratories. The IG Farben plant in Bitterfeld cooperated, for example, with the physicist Adolf Smekal at the University of Halle on the visualisation of scratches on glass surfaces (Th. Marx, W. Klemm and A. Smekal, 'Übermikroskopische Struktur von Ritzbahnen', *Die Naturwissenschaften* 11 (1943), 143–4).

52 C. Ramsauer (ed.), *Zehn Jahre Elektronenmikroskopie* (Berlin: Springer, 1941); C. Ramsauer (ed.), *Elektronenmikroskopie. Bericht über die Arbeiten des AEG Forschungs-Instituts, 1930–1941* (Berlin: Springer, 1942), and C. Ramsauer (ed.), *Elektronenmikroskopie. Bericht über die Arbeiten des AEG Forschungs-Instituts, 1930–1942* (Berlin: Springer, 1943).

53 M. von Ardenne and D. Beischer, 'Untersuchung von Katalysatoren mit dem Universal-Elektronenmikroskop', *Angewandte Chemie* 53 (1940), 103–7. On his collaboration with scientists from the Kaiser Wilhelm Institut for Biochemistry, see e.g. M. von Ardenne, H. Friedrich-Freksa and G. Schramm, 'Elektronenmikroskopische Untersuchung der Präcipitinreaktion von Tabakmosaikvirus mit Kaninchenantiserum', *Archiv für die Gesamte Virusforschung* 2 (1941), 80–6.

54 A popular innovation was the introduction of a cine camera and cinematographic methods in electron microscopy (M. von Ardenne, 'Elektronenmikrokinematographie f authorslaboratories at Siemens mit dem Universal-Elektronenmikroskop' *Zeitschrift für Physik* 120 (1943), 397–412).

55 Note, 27 October 1942; SAA 11 Flir Lg 177.

56 B. von Borries, 'Sublichtmikroskopische Auflösung bei der Abbildung von Oberflächen im Übermikroskop' *Zeitschrift für Physik* 116 (1940), 370–8 and B. von Borries and W. Ruttmann, 'Metallographische Untersuchungen mit dem Übermikroskop an Stahl, Gußeisen und Messing', *Wissenschaftl. Veröffentlichungen aus den Siemens-Werken*, (Berlin: Springer, 1940), 342–62.

57 BIOS Nr. 1671(ref. 8), p. 5.

58 Copy of a letter to Dr Hertz at the Hauptausschuss 'Nachrichtengerät', Reich Ministry for Armament and Ammunition, 22 October 1942, SAA 11 Flir / Lg 177.

59 For these purposes a super microscope was delivered to Carl Zeiss Jena in 1941 (Wolpers 1991, 224).

60 See e.g., B. von Borries and S. Janzen, 'Abbildung feinbearbeiteter technischer Oberflächen im Übermikroskop', *VDI-Zeitschrift* 85 (1941), 207–11.

61 Copy of a letter to Dr Hertz. The last entry probably refers to research being done at Strasbourg University under the supervision of the professors of anatomy, August Hirt, and biology, Otto Bickenbach. Hirt and Bickenbach conducted research into the impact of mustard gas and phosgene gas on the human organism (Hirt to Wolfram Sievers, 29

March 1943; BA-Berlin, NS 21 'SS-Ahnenerbe' / 906; on Bickenbach's research, see F. Schmaltz, 'Otto Bickenbach's Human Experiments with Chemical Warfare Agents and the Concentration Camp Natzweiler', W. U. Eckart (ed.), *Man, Medicine, and the State* (Stuttgart: Franz Steiner, 2006), 139–56: 145).

62 H. Fries and H. O. Müller, 'Staube und Rauche im Übermikroskop', *Die Gasmaske. Zeitschrift für Atemschutz* 11 (1939), 1–9.

63 E. Geißler, *'Biologische Waffen – nicht in Hitlers Arsenalen. Biologische und Toxin-Kampfmittel in Deutschland von 1915 bis 1945'* (Münster: Litt, 1999), 606–11; G. Riedel and H. Ruska, 'Übermikroskopische Bestimmung der Teilchenzahl eines Sols über dessen aerodispersiven Zustand', *Kolloid-Zeitschrift* 96 (1941), 86–96; G. Riedel, 'Ein elektrischer Kernfäller zur Gewinnung übermikroskopischer Präparate', *Kolloid-Zeitschrift* 103 (1943), 228–32; from these publications, which mention the extreme dangerousness of some of the substances used in the experiments, it is not clear if the exact background of these experiments was known to his colleagues at Siemens.

64 Ernst Ruska to Helmut Ruska, 16 June 1942, MPG Archive, III, ZA 137 / 18.

65 A. Tiselius and S. Gard, 'Übermikroskopische Beobachtungen an Poliomyelitis-Viruspräparaten, *Die Naturwissenschaften* 49/49 (1942), 728–31; H. Ruska, 'Der Erreger der spinalen Kinderlähmung', *Umschau* 47 (1943), 216–17. 'While several weeks ago a guest of our laboratory, a Swedish professor, managed to visualise the polio virus, we expect a large Italian group in the next week' (Copy of a letter from Flir and v. Borries to Generaloberststabsarzt Siegfried Handloser, 29 September 1942, MPG Archive, III, ZA 137 /18).

66 After Ernst Ruska had left the Technical University Driese, Müller, Krause and Beischer used his instrument for their own research (improving its performance at some point); see E. Drieste and H. O. Müller, 'Elektronenmikroskopische Aufnahmen (Elektronenmikrogramme) von Chitinobjekten', *Zeitschr. Wiss. Mikroskopie* 52 (1935), 53–7; F. Krause, 'Elektronenmikroskopische Aufnahmen von Diatomeen mit dem magnetischen Elektronenmikroskop', *Zeitschrift für Physik* 102 (1936), 417–22.

67 Helmut, Ernst Ruska and v. Borries in a letter to Conti, 7 March 1941: 'Vorschlag zur Gründung eines Reichsinstituts für medizinische Strukturforschung', Archive Institut für Stadtgeschichte, Frankfurt am Main, Magistratsakten, 8.629.

68 A critical account of the value of the application of electron microscopy in the medical and biological sciences can be found in the Fiat Review on biophysics: G. Piekarski, 'Das Elektronemikroskop in Biologie und Medizin', B. Rajewksi and M. Schön (eds), *FIAT-Review of German Science*, 1939–1946, *Biophysics, part II* (Wiesbaden: Dieterichsche Verlagsbuchhandlung, 1948), 173–209: 209.

69 von Buol to Lüschen, 19 October 1942 (transcription); SAA 11 Flir / Lg 177.

70 Bodo v. Borries to von Buol, 23 February 1939; SAA 11 Flir / Lg 177.

71 Ibid.

72 In January 1943, Morell recalled a visit to the Siemens laboratory 'several years ago' commemorating Hitler's promise to provide him with an electron microscope (Morell to v. Borries and Ernst Ruska, 25 January 1943, MPG Archive, III, Za 137/ 18). Erhard Geißler reports that in May 1944 Hitler ordered the delivery of an instrument to Morell, which had originally been ordered by the Military Medical Academy (Geißler 1999 [ref. 63], 611).

73 R. Bülow, 'Das Übermikroskop aus dem Deutschlandhaus', *Kultur & Technik* 1 (1996), 31–4.

74 H. Lammers (head of the Reichskanzlei) to Mahl, 29 May 1943, Staatsbibliothek Preussischer Kulturbesitz, Berlin, Nachlass Hans Mahl.

75 'Verfügung über die Stiftung des Fritz-Todt-Preises', 2 February 1944, in: Max Domarus, *Hitler. Reden und Proklamationen*, 1932–1945, Volume II, (Würzburg: Domarus Verlag, 1963, 2087).

76 Bodo von Borries, 'Über die geschäftlichen Aussichten der Übermikroskopie', copy of a memo, 2 February 1947, MPG Archive, III, 137 / 42.
77 'A few other American firms [apart from RCA], such as Kodak and GE, built electron microscopes in the late 1930s, but none was very serious about commercialising them, believing that even a handful might saturate the market' (Rasmussen 1997, 3).
78 Ibid.
79 '[I]n the context of wartime propaganda the electron microscope was essentially significant as an unclassified high-technology development with which to counter German claims of scientific superiority'. (Rasmussen 1997, 51).
80 Rasmussen 1997, 227.
81 H. Ruska, 'Onderzoekingsmethoden en Resultaten der Supermicroscopie' *Nederlandsch Tijdschrift voor Natuurkunde* 7 (1940), 179–91. On the visit of members of the Siemens team to the laboratory of J. B. le Poole in Delft, see the paper by Marian Fournier in this volume.
82 H. Eyer and H. Ruska, 'Über den Feinbau der Fleckfieber-Rickettsie', *Zeitschrift f. Hygiene und Infektionskrankheiten* 125 (1944), 483–92: 483.
83 E. Brüche, 'Das Elektronemikroskop', *Die Donaubrücke. Organ der Deutsch-Ungarischen Handelskammer* 1 (1944), 16–17.
84 Helmut Ruska, Report on a lecture tour to France, undated; SAA Flir 11 / Lg. 29.
85 Ibid. The bibliography of papers published in association with the Laboratories of Electron Optics and Super Microscopy (*Veröffentlichungen aus den Laboratorien für Elektronenoptik und Übermikroskopie der Siemens & Halske Ag, 1938–1949*, Berlin: Siemens & Halske, undated) lists three publications of Levaditi and Bonét-Maury; I could not, however, find out if they actually stayed as guests in Berlin; see C. Levaditi, 'Aspect et dimensions des corps élémentaires vaccinaux et des corpuscules normaux en lumière électronique', *Monographies de l'Institut Alfred-Fournier* (1943), 1–4; P. Bonét-Maury, 'La mesure des dimensions du virus vaccinal d'après la micrographie électronique', ibid., 5–6; C. Levaditi 'Le Treponema palladium en microscopie électronique', ibid., 7–10.
86 Helmut Ruska, Report on a lecture tour to France, undated; SAA Flir 11 / Lg. 29.
87 Rostock to Mentzel, 2 November 1944, BA-Berlin, R 26 III / 199.
88 von Borries and Ernst Ruska to Rostock, 21 November 1944, MPG Archive, III, ZA 137 /18.
89 Rostock to Mentzel, 2 November 1944, BA-Berlin, R 26 III / 199.
90 L. Kühle, 'Werden Moleküle sichtbar? Glänzende Entwicklung der Elektronenmikroskopie', *Völkischer Beobachter*, 2 December 1944.
91 H. Mahl, 'Metallkundliche Untersuchungen mit dem Übermikroskop', *Zeitschrift für technische Physik* 21 (1940), 17–18; H. Mahl, 'Die übermikroskopische Oberflächendarstellung mit dem Abdruckverfahren', *Die Naturwissenschaften* 14/15 (1942), 207–17.
92 'Report on the status of the development of electron microscopy', 7 December 1944, LTA, Brüche II / 279.
93 Brüche to Joos, 22 January 1945, Carl Zeiss Archive, BACZ 23423.
94 Ramsauer to Zenneck, 9 January 1943, LTA, Brüche II / 279.
95 Summary of correspondence between Zeiss and AEG and of the inspection report, LTA, Brüche II / 279.
96 Georg Joos to Brüche, 1 December 1944, LTA, Brüche II / 279.
97 'Durch Erlaß des Führers vom 31.1.45 sind alle Arbeiten der Rüstungsfertigung unbedingt fortzusetzen. Dementsprechend erhält hiermit die Arbeitsgruppe Übermikroskopie der Firma Siemens unter der Leitung der Herren v. Borries und Dr. Ruska die Weisung, die laufenden Arbeiten unbedingt fortzusetzen, da die Geräte schnellstens benötigt werden'. Graue, Leiter der Kriegswirtschaftstelle beim Präsidium des Reichsforschungsrates, 5 February 1945; MPG Archive, III, ZA 137 / 42. In October 1944,

twenty-eight electron microscopes were ordered; seven had the highest urgency level 0 DE, one SS I, three SS II, six SS III and 11 lesser levels or none (Internal report on the institute's development in 1943/44, MPG Archive, III, ZA 137 / 42).

98 Geißler 1999 (ref. 63), 599.

99 'Electron Microscopy in Germany', BIOS. 1671, 10.

9 Aerodynamic research at the Nationaal Luchtvaartlaboratorium (NLL) in Amsterdam under German occupation during World War II

Florian Schmaltz

Shortly after the invasion of The Netherlands by Nazi Germany in May 1940 the most important Dutch aerodynamic research establishment, the Nationaal Luchtvaartlaboratorium (NLL) in Amsterdam, came under the administrative control of the Aerodynamische Versuchsanstalt Göttingen (AVA). What kind of impact did the occupation have on the development of the NLL? In what ways did Dutch scientists contribute to the war effort of Nazi Germany? How did the occupation policy affect the economic development of the NLL, its staff and scientific research? What strategies were used by the occupying power to make the Dutch scientists collaborate? How did the German scientists ensure that war-relevant knowledge was kept secret? Was there any resistance? Are there specific periods to be identified in the relations of Germans scientists in charge of the NLL and with regard to the repression of resistance?

This paper begins with a brief overview of the institutional development of the Aerodynamische Versuchsanstalt (AVA) in Göttingen and its satellite institutes in countries occupied by Nazi Germany. A description of the resources and wind tunnels of the NLL in Amsterdam made available to the AVA is given, followed by some information about the economic development of the NLL. It then goes on to analyse examples of military research conducted at the NLL for Nazi Germany. It concludes with a discussion of the crisis in occupation policy in 1943 and 1944, with the tightening of repression, and the growing importance of the NLL as a social organisation for its staff in surviving the hunger winter of 1944/45.

The AVA and its International Network of Satellite Institutes in German-occupied Territories during World War II

After the Nazis had seized power in 1933, aeronautical research boomed in Germany. In defiance of the Treaty of Versailles the new armament policy expanded the annual budgets and staff of all German aeronautical institutes.[1] The predecessor of the AVA in Göttingen, originally founded as the Modellversuchsanstalt (Model Experimental Institute) in 1907, did not possess a large wind tunnel before the First World War. With the support of the War Ministry new buildings and a new and larger wind tunnel were constructed in Göttingen. The growing importance of aerial warfare led to a tremendous expansion of the facilities owing to the requirements of the military.[2] Renamed the Aerodynamische Versuchsanstalt in 1920, the Göttingen institute survived the First World War but suffered severe cuts to its annual budgets during the days of the Weimar Republic. In July 1925, the Kaiser Wilhelm Institute for Fluid Dynamics was opened in conjunction with the AVA in Göttingen.[3]

Only weeks after Hitler was appointed Chancellor in 1933, the institutes in Göttingen received 2.5 million Reichsmark to expand their facilities and construct a large wind tunnel.[4] The staff of the AVA grew from 80 employees in 1933 to more than 530[5] by 1938 and reached more than 700 by 1940.[6] In April 1937, the AVA was separated from the Kaiser Wilhelm Institute for Fluid Dynamics and re-founded as a registered corporation. One of the effects of this was that the Kaiser Wilhelm Society was relieved of the burden of the greatly increased administrative work in connection with the growing staff.[7] Two years later, in 1939, the AVA was reorganised once again and divided into eight institutes with a specific division of labour (Table 9.1).

The occupation of foreign countries by Nazi Germany allowed the AVA to expand further and opened up new perspectives for its scientists. During World War II a widespread network of satellite institutes controlled by the AVA was established (Table 9.2).

Immediately after the invasion of Western Europe in May 1940 several large aerodynamic institutes in France and The Netherlands came under the control of the AVA. In France, the Institute Aérotechnique de Saint-Cyr and the wind tunnels of Hispano Suiza (Paris) were among the institutes seized.[8]

From the summer of 1940 onwards the AVA established a satellite institute in Prague ('Protectorate of Bohemia-Moravia') for icing experiments.[9] In 1941, the AVA built facilities in the mountains at Kufstein in Austria to test the destruction of propellers using steel cables. In the same year the AVA established an open-air

Table 9.1 Institutes of the AVA after 1939[10]

Institute for	*Director*
Wind Tunnels	Prof. Reinhold Seiferth
Flight Testing and Aviation	Prof. Josef Stüper
Icing Research	Dipl.-Ing. Ludolf Ritz
Theoretical Aerodynamics	Prof. Albert Betz
Turbo-machines	Dipl.-Ing. Walter Encke
Unsteady Fluid Motion	Prof. Hans-Georg Küssner
High Speed Research	Prof. Otto Walchner
Instrument Development	Dipl-Ing. Otto Mühlhäuser

Table 9.2 'Satellite institutes' of the AVA in occupied countries during World War II

Period	*Location/Institution*	*Head*	*Main research topics*
1941 – (?)	Kufstein/Eibergkopf (Austria)	Dipl. Ing. Otto Mühlhäuser	Destruction experiments with steel cables
(Sept.?) 1940 – April 1945	Prague – Institute for Icing Research ('Protectorate of Bohemia-Moravia')	Dipl. Ing. Helmut Glaser	De-icing research on aircraft; development of flight instruments
May 1940 – Sept. 1944	*Institute Aérotechnique de Saint-Cyr* and *Hispano Suiza* (Paris)	Dipl. Ing. Josef Käufl (Prof. Dr. Albert Toussaint)	Wind tunnel experiments Theoretical aerodynamics
May 1940 – Sept. 1944	*Nationaal Luchtvaartlaboratorium* (Amsterdam)	Dipl. Ing. Josef Käufl (Dr. Ing. Carel Koning)	Wind tunnel experiments Theoretical aerodynamics
1943 – 1944	Pernau near Riga (Latvia)	Dipl. Ing. Karl Schlör	Aerodynamics and de-icing of snowmobiles
1941 – (?)	Finse (Norway)	Dr. Ing. Franz Eder (Dipl. Ing. Karl Schlör)	De-icing in open-air test unit
Nov. 1941 – Feb. 1943	Charkov (Aerodynamic Institute), evacuated to Ummendorf (Württemberg) continued in November 1943	Prof. Dr. Dipl. Ing. Michael Strscheletzky	Development of measurement methods; cavitation tests and propeller research

test unit for de-icing experiments at Finse in Norway.[11] After the invasion by the *Wehrmacht* (German armed forces) in Eastern Europe the Aerodynamic Institute of the Technical University of Charkov came under the control of the AVA in November 1941 and was utilised for theoretical and experimental studies of cavitation. Two years later, this institute was evacuated to Ummendorf (Württemberg) in southern Germany, but remained under the supervision of the AVA.[12] In Latvia, AVA researcher Karl Schlör conducted aerodynamic experiments at an outpost of the AVA near Riga to improve the design of snowmobiles built for military purposes in winter warfare.[13]

Josef Käufl, group leader of the AVA's large wind tunnel number VI in Göttingen, became responsible for the administration of aeronautical institutes in the Western occupied territories in the summer of 1940. Käufl, born in Landshut (Bavaria) in 1907, had studied mechanical engineering at the Technical University of Munich, where he received his diploma as an engineer in 1930. Having finished university, he worked for Robert Bosch AG, Landshut, from May 1930 until July 1931. He then moved to Tilburg in The Netherlands,

Figure 9.1 Josef Käufl.
Source: DLR Archives, Göttingen

where he became a foreign representative for a Bosch agency for four years. In February 1935, Käufl joined the AVA in Göttingen as a scientific assistant. Initially, he worked with the medium-sized wind tunnel and from August 1936 onwards with the large wind tunnel number VI. In January 1940, he became group leader of this tunnel and a year later was appointed head of department.[14] In addition to his professional skills, his ability to speak Dutch was probably one of the reasons why he was selected to take over the post as official representative of the liaison departments of the *Generalluftzeugmeister* (General Aircraft Inspector) in Paris and in Amsterdam established in 1940.[15]

Resources, Wind Tunnels and Research Profile

Just a week after the German invasion of The Netherlands, France, Belgium and Luxembourg, on 17 May 1940, Josef Käufl received orders to survey aeronautical establishments in the western occupied territories. To organise the exploitation of scientific resources for the German war effort Käufl was armed with a gun during his inspection trip.[16] By the end of May the *Luftgaukommando* Holland was informed by the German Air Ministry that Käufl had been authorised to continue business and 'changes and expansion of the experimental installations were, for the time being, only to be undertaken with his allowance'.[17] The detailed intelligence included in his first report to the AVA in Göttingen, dating from early June 1940, shows that he obtained access to the records of the NLL in Amsterdam. This first report delivered information about the legal status of the NLL, its financial and organisational background and the technical facilities available. Without mentioning him by name, Käufl noted that the present director of the NLL, Emile Benjamin Wolff, was very ill and had asked for his resignation.[18] In August 1940, the board of the NLL appointed Carel Koning as scientific director and Jean Louis Chaillet as commercial director.[19] It is unclear if the resignation of Wolff was motivated by his illness alone. Wolff was Jewish and for that reason it was clear that it was only a matter of time before the German occupiers would have removed him from his post. In the course of the persecution of the Jews, in November 1940, soon after his resignation, all Jewish civil servants in The Netherlands were dismissed.[20] Wolff died in Amsterdam on 7 February 1941, a month after the deportations of the Jews from The Netherlands to the concentration and destruction camps had begun.[21]

The NLL was organised into four main departments (Table 9.3): the Aerodynamics Department, the Structure Department, the Aircraft Department and the Material Research Department. There was also a workshop for instrument construction, as well as a Mechanical Department, a model joinery and an editorial office for the compilation of reports. Finally, the NLL had a Department for General Services and another Department for Administration. In 1940, several changes occurred concerning the heads of the departments. When Carel Koning took over the post of scientific director in October 1940, Adrianus Boelen became his successor as head of the Aerodynamics Department. The head of the Aircraft Department, Hendricus Jacobus van der Maas, left the NLL and became professor at the Technical University of Delft on 6 May 1940. His successor was Anthonie Jacob Marx. The other heads of department remained unchanged. Arie van der

Table 9.3 Departments of the Nationaal Luchtvaartlaboratorium (NLL) Amsterdam in 1940[22]

Aerodynamics Department (Aerodynamische Afdeeling)	*Structure Department (Sterkte Afdeeling)*	*Aircraft Department (Vliegtuigen Afdeeling)*	*Material Research Department (Materialen Afdeeling)*
Carel Koning (until September 1940) Adrianus Boelen (from 1 Oct 1940)	Arie van der Neut	Hendricus Jacobus van der Maas (until 6 May 1940) Anthonie Jacob Marx (from May 1940)	Leopold Johan Gerhard van Ewijk

Neut was still in charge of the Structure Department and Leopold Johan Gerhard van Ewijk continued his work as head of the Material Research Department.[23]

When the NLL fell into German hands, the laboratory had just moved to a new building that possessed two large, modern wind tunnels.[24] In June 1940, Käufl reported to the AVA and the German Air Ministry that the NLL's large wind tunnel had a jet of 2.1 x 3 m, and was 'ready for use', whereas the second tunnel, with a smaller diameter of 1.5 x 1.5 m, still had 'to be completed with the appropriate power engine'. Käufl added:

> The old tunnel, situated at the Navy Yard, has been dismantled and the machines sent in for overhauling. The buildings of the new institute, which are presently furnished and fettled, are located at Sloterweg, about 3 kilometres away from Schiphol airport. When the buildings are finished, at first the large wind tunnel will be put in operation and the beginning of operation is assumed for 1 July.[25]

Figure 9.2 The large wind tunnel at the NLL Amsterdam.
Source: Stichting Historisch Museum NLR, Amsterdam

Käufl attached detailed technical drafts to his report about the tunnels, both of them capable of operating with a closed and an open measuring section. He ended his report with descriptions of the equipment in the workshops and mentioned the possibility of taking up flight tests with a Fokker F7 owned by the NLL at Schiphol as soon as the military occupation allowed the resumption of air traffic at the airport.[26] Clearly, the Germans were aware of NLL's laboratory aircraft. Therefore, the claim by Jan van der Bliek in his book about the history of the NLL that the aircraft had been hidden by NLL employees during the occupation period at the Van Dam yacht harbour, Oude Wetering, is improbable.[27] Both wind tunnels at the NLL went into operation in 1940. The larger tunnel started operation on 17 June 1940 – one month after Käufl had visited the NLL – and the second largest started operation on 20 November 1940.[28] In addition to the two larger wind tunnels the NLL also possessed a small model wind tunnel (no. 2), which was an exact replica of the large tunnel built on a reduced scale of 1:10 to examine the properties of its design before the construction of the large tunnel (Table 9.4).[29]

Table 9.4 Wind tunnels at the NLL in Amsterdam[30]

Name	*Built*	*In operation*	*Operated by*	*Diameter (metres)*	*Speed (metres/ sec.)*
No. 1 *Oude Windtunnel* (Old Wind Tunnel)	1918	1918–mid-1940	RSL and NLL	1.60	30
No. 2 *Model Windtunnel* (Model Wind Tunnel) (scale 1:10 of No. 3)	1939–1940	?	NLL	0.21 x 0.30 (0.30 x 0.40)	73 (35)
No. 3 *Groote Windtunnel* (Large Wind Tunnel)	1939–1940	1940	NLL	2.10 x 3.03 (3.00 x 4.00)	81 (36)
No. 4 *Kleine Windtunnel* (Small Wind Tunnel)	1939–1940	1940	NLL	1.51 x 1.51 (2.00 x 2.00)	40 (23)

Before the end of July 1940 a delegation from AVA including the scientific director, Albert Betz, and Josef Käufl went on a round trip to inspect aeronautical research facilities in Amsterdam, Wageningen, Brussels, Lille and Paris. Where Amsterdam was concerned, the delegation came to the following conclusion:

> The facilities consistently make a very thought-out impression and appear to have a significant performance capacity. The installations themselves do not go beyond what is known in Germany. The instrumentation had to be modified because of the limited funds and therefore does not meet the technological standards in Germany. The most valuable part of this establishment is its staff, especially the high quality scientific capacity. The utilisation for German interests would therefore be achieved most conveniently by placing orders for specific research work. This type of utilisation would be facilitated because the head of the establishment, the engineer Mr. Koning, obviously shows a quite accommodating attitude towards Germany.[31]

In August 1940, Albert Betz submitted a research programme to the German Air Ministry, which included a list of aerodynamic topics to be conducted in Amsterdam 'without any concerns'.[32] Six of the seven research projects Betz had suggested dealt with aircraft wings, one with propellers. The research programme of wind tunnel tests included measurements of the following aircraft parts:

1. the wing fuselage area, to avoid early airfoil stalling without producing a substantial increase in drag in high-speed flight;
2. wings with flaps to define the increase in lift for its application as a starting and landing aid;
3. trapeze wings, protected against stalling by stronger curved profiles;
4. the ground influence on wings of maximum lift;
5. wings with Betz slots and wings with fixed slots on the outboard wing (it was expected that both types of slots would improve tailspin performance);
6. the pressure distribution at tails with rudder gaps; and
7. theoretical studies of the influence of propeller jets on the lift distribution of a wing in the jet.[33]

This last research project aimed to improve the stability and steerageway of the aircraft. For this purpose, NLL director Koning had developed a theory that allowed calculation of such problems for the first time.[34] The research programme was approved by the German Air Ministry a few days later.[35]

On 2 September 1940, the German Air Ministry assigned the AVA the task of exclusively handling all research projects given to the NLL by any German aeronautical establishments or German aeronautical company.[36] At the end of the same month a conference was convened in Amsterdam, at which representatives of the German occupation authorities and the Dutch administration, including representatives from the Dutch Ministry of Interior Affairs (*Departmement van Binnenlandsche Zaken*) and the Department van of Water Management (*Department van*

Waterstaat) took part. The German Air Ministry was authorised to be 'informed about all works and projects' of the NLL and attend all its meetings. New rules were instituted to secure secrecy for German research projects, while the question of how to finance the NLL remained unresolved.[37] In mid-December 1940, the German Air Ministry approached the General Commissioner for Administration and Justice to effect a change in the articles of the NLL to ensure secrecy of German research from Dutch ministries and to allow German authorities to control the supervisory board of the NLL.[38] The Stichting (Foundation) NLL accepted the new articles with some minor changes.[39]

Research at the NLL for the AVA in Göttingen

Up until the German occupation the Material Research Department had contributed to the review work of the regulations concerning aircraft materials of the Dutch Air Force.[40] This work was stopped immediately after the German invasion. However, the AVA showed very little interest in the Material Research Department. As far as available documents show, this department only produced four research reports for the AVA during the whole period of German occupation.[41] More time was devoted to theoretical studies, experiments about corrosion of aircraft paint, tests of new alloys and bolted fastenings.[42] In 1941, reports of earlier experiments were expanded and the construction of a machine for tests on bolted fastenings was completed.[43] The department continued with research and the development of instruments, but encountered difficulties when, from 13 June to 2 September 1942, Head of Department van Ewijk was detained for unknown reasons.[44]

The Structure Department carried out some research work for the AVA. Up until May 1940, it had been involved in the work of a commission to develop the statics regulations for civil and military aircraft. The department developed a mathematical method for the approximation of pressure distribution transverse along the wing.[45] Some of the theoretical work of this department conducted for Germany by Arie van der Neut in 1942 might have been connected with problems of stability of ballistic rockets.[46] Other research projects, conducted for the AVA by Izak Brinkhorst from 1941 until 1943, dealt with torsion stiffness of aircraft wings.[47] In 1944, Arie van der Neut finished two reports about torsion experiments of aircraft wings in which theory and experiments were combined.[48]

More important for the AVA was the Aircraft Department, even though its research practice underwent a profound change when German occupation authorities prohibited Dutch pilots from conducting flight tests. After the suspension of experimental flights, the department continued its work by switching completely to theoretical studies. According to the NLL annual report of 1940, a significant part of the research work of the Aircraft Department came from the military and was therefore cancelled after the Dutch capitulation. Some work time was consumed by the installation of the laboratories in the new buildings and for calibration of the new wind tunnels.[49] The theoretical research included studies of numerical methods for the calculations of wing flutter for Hans-Georg Küssner of the Institute for Unsteady Motions of the AVA in Göttingen.[50] In 1941, further wing flutter experiments for Küssner were prepared by Johan H. Greidanus.[51] In the following year a gyroscope was tested for the AVA.[52] In the autumn of 1942 and the spring of 1943, Greidanus and van de Vooren compared the mathematical results of a theoretical study of wing flutter with results achieved in experimental studies in Germany. The Dutch scientists were not informed that the wings examined were from a Junkers 52 aircraft.[53] In 1943 and 1944, several studies were conducted to improve instruments for experiments and measurements, such as variometers[54] and gyroscopes.[55]

From the German perspective, the Aerodynamics Department was the most important department at the NLL. Even though most of the work during the early months of the occupation dealt with the installation of the laboratory in the new building and the calibration of the new wind tunnels, no other department produced as many studies for the AVA. In 1940, the Aerodynamic Department was busy evaluating the characteristics of the jet in the new wind tunnels. The data from those measurements were an essential precondition to taking over research projects.[56] In the first half of 1940, ongoing work was also slowed down because the head of the department, Carel Koning, was acting as the representative of the director Emile Wolff, who was unable to fulfil his duties owing to his illness.[57] In order to catch up with the research projects, a double shift of 16 hours was introduced in the large wind tunnel in April 1941.[58] Moreover, the workflow was improved by better coordination of assembly work, calibration of the balances, and measurements.[59] In the second half of 1943, the Aerodynamic Department was confronted with the problem of qualified employees being recruited as forced labourers in Germany. Younger workers, who had to become acquainted with the research work, were hired to fill the gap.[60]

Among the first reports translated for the AVA in 1941 were studies about model measurements of aeroplane skis with built-in wheels originally conducted for a Dutch company by the Aerodynamic Department of the NLL.[61] These studies were continued the following year with other ski model types.[62] The first report on wind tunnel tests conducted for the AVA in 1941 (which was, in fact, a camouflaged research order by the Messerschmitt Company) dealt with the drag and airstream through cooling units attached to a wing.[63] Further studies for the AVA in 1942 tested the quality of the jet in the new large NLL wind tunnel (no. 3).[64] In May 1942, Koning finished a theoretical study for the AVA into the wall influence in propeller measurements in wind tunnels and possible mathematical corrections.[65] In August 1943, the wall influence of the smaller wind tunnel (no. 4) was tested by the Aerodynamic Department with two different nozzles and varying shapes of the measuring section.[66]

In the same year, the Aerodynamic Department undertook a series of studies for the German aircraft company Arado. Until May 1943, a 1:10 model of the Arado Ar 233 aircraft without working propellers and the propellers were tested separately.[67] The Arado Ar 233, designed by Arado as a civilian seaplane, and tested by the French company Dewoitine in 1942, was ordered by the German Air Ministry in 1940. Finally, in 1943, the Arado Ar 233 model was tested at the NLL in

Amsterdam with running propellers.[68] Completion of a mock-up Ar 233 remained unfinished in 1944 because of the war.[69] In the same year, the Aerodynamic Department undertook two studies with an NACA profile with flaps to improve lift during starting and landing.[70] Another topic investigated in two studies between 1943 and 1944 entailed measurements of rectangular wings with propeller brakes in an open circuit.[71]

Another project linked to military research concerned tests on eight types of swept-back wings conducted for the Messerschmitt Company.[72] Among the last research orders from the AVA in Göttingen, finished in 1944, were wind tunnel measurements of wing models with landing flaps.[73]

Technical Difficulties with the Wind Tunnels of the NLL

After the board of the Stichting NLL had come to the decision to construct a new laboratory with new wind tunnels in 1938, Carel Koning had visited the AVA in Göttingen in August of the same year to gather information about the latest developments in wind tunnel construction.[74] However, the exchange Koning established with the director of the Aerodynamic Institute of the ETH in Zurich, Jakob Ackeret, was clearly more important.[75] Ackeret, who was a pupil of the aerodynamicist and director of the Kaiser Wilhelm Institute for Fluid Dynamics in Göttingen, Ludwig Prandtl, became a consultant for the NLL when the new laboratory's wind tunnels were being planned. Finally, the ETH Zurich wind tunnel became the model for the large NLL wind tunnel. However, the tunnel in Amsterdam was not an exact copy since its length and its contraction were changed to achieve a lower degree of turbulence.[76] Another difference, compared to the Swiss tunnel, was the tunnel's drive system. Instead of two propellers as in Zurich, the tunnel in Amsterdam had just one large, six-bladed propeller with a diameter of 4.20 m.[77] This construction, designed by Professor Johannes Martinus (Jan) Burgers of the Technical University of Delft, proved to be the weak point in the new experimental apparatus.[78] Having been in operation for 14 months, the propeller blades began to cause a series of difficulties. On 10 September 1941, one of the blades in the large wind tunnel broke for the first time.[79] The large wind tunnel was out of action until at least the end of 1941, if not longer.[80] During this long delay the Aerodynamics Department focused on the improvement of the six-component balance as an important instrument for the wind tunnel measurements, experimented with the model wind tunnel (no. 2) and a provisional high-speed tunnel.[81] Because of supply shortages it took six months before the new blades made from hydronalium (an aluminium-magnesium alloy) were delivered. In the meantime, three blades of the old wind tunnel were installed as a provisional solution. To avoid any recurrence of the damage, the maximum operating speed of the wind was limited to an air speed of 62 metres per second, which was seen as an unacceptable condition by the Dutch engineers.[82] But on 23 March 1942, a crack was discovered in one of the 11 flow blades positioned in front of the large wind tunnel's propeller. The next day the material fatigue continued and two more blades showed an additional three cracks. Koning informed the AVA that he expected the large wind tunnel to come to a complete standstill because of this damage. The NLL director estimated three weeks for the repair work, if the necessary material for strengthening the blades were available.[83] On 13 April 1942, shortly after the large wind tunnel went into operation again, one of the blades of the large wind tunnel broke.[84] During a meeting of the board of the Stichting NLL Koning presented a short report about the difficulties with the blades since 1940. In their fault analysis, the Dutch scientists followed the hypothesis that the damage observed was connected with the aerodynamic problem of reaching a continuous air stream in the wind tunnel. According to their explanation of why the blade fatigue occurred, vibrations were caused by the characteristics of the air stream at certain speeds.[85]

Five months later, on 20 October 1942, Koning had to inform the AVA that the blade of the large wind tunnel had been broken again. He expected a further delay of at least another two months.[86] From 1943 onwards, no further difficulties with the propeller blades of the large wind tunnel were reported.

Was the repeated damage to the blades the result of active sabotage to foil effective work for the German occupier? Even though we cannot exclude this possibility, it seems unlikely that this was the reason for the difficulties. Since the administration of the NLL had a strong financial interest in continuing operation in order to guarantee revenues from German and Dutch research orders, active sabotage would have endangered the resources of the laboratory. Less obvious and therefore less dangerous forms of sabotage were possible by manipulating the data of the measurements slightly, making the results worthless for the German side. Adriaan de Lathouder of the Aerodynamic Department claimed this form of resistance was practised at the NLL.[87] However, it is difficult to find hard evidence of this claim in the sources since the original data of the measurements are no longer available.

Economic Development of the NLL

With regard to the economic occupation policy of Nazi Germany towards The Netherlands, the historian Gerhard Hirschfeld has remarked that the German authorities established indirect economic control in Western Europe and 'left day-to-day economic affairs in native hands whilst keeping overall supervision and direction in the hands of the occupation administration'.[88] Can we find similar patterns in the occupation policy concerning the exploitation of scientific resources?

In May 1940, during his first visit to Amsterdam, Käufl had studied the annual reports and received internal information about the financing of the NLL's budget. The legal constitution of the NLL as a Stichting, he wrote in his first report on the NLL, had existed since 1937 and corresponded to the German foundation. The NLL was financed in part by contracts received by the industry and in part by public support from the Nederlandse Centrale Organisatie voor Toegepast – Natuurwetenschappelijk Onderzoek (TNO). The members of the TNO enjoyed price reductions from the individual institutes when they gave them orders in the members' own private interests. The NLL, for example, charged members a surcharge on costs of 150 per cent and non-members one of 200 per cent.[89]

The German occupiers agreed to pay the regular price for research contracts. They took advantage of the fact that the TNO and Dutch ministries continued to finance the general expenses for NLL infrastructure and staff. This interest met with the efforts of the Stichting NLL to maintain service of the laboratory, secure its future existence and keep as much control as possible over the institution in Dutch hands. This solution was in the interest of both sides. On the German side, neither the Air Ministry nor the AVA in Göttingen had to meet any costs for the maintenance of the laboratory's infrastructure and the employment of the NLL staff. On the Dutch side, continued financing secured a certain control over the laboratory.

Development of the Annual Budget of the NLL

How did the annual budget of the NLL develop under German occupation? Compared to the annual expenses of the pre-war period, the figures during the first year of the occupation had risen by no more than about 10 per cent from Hfl. 231,296 in 1939[90] to Hfl. 254,073 in 1940.[91] In 1941, the new wind tunnels were in operation and expenses for staff and other costs rose by more than 30 per cent from Hfl. 329,685 in 1941[92] to 433,939 in 1942.[93] After rising slightly to Hfl. 443,842 in 1943,[94] the annual budget reached its peak in 1944 at Hfl. 498,297.[95] The figures show that the expenses of the NLL more than doubled between 1940 and 1944. In contrast to the devastating economic and financial plundering of The Netherlands during the German occupation, the microeconomic development of the business of the NLL was booming.

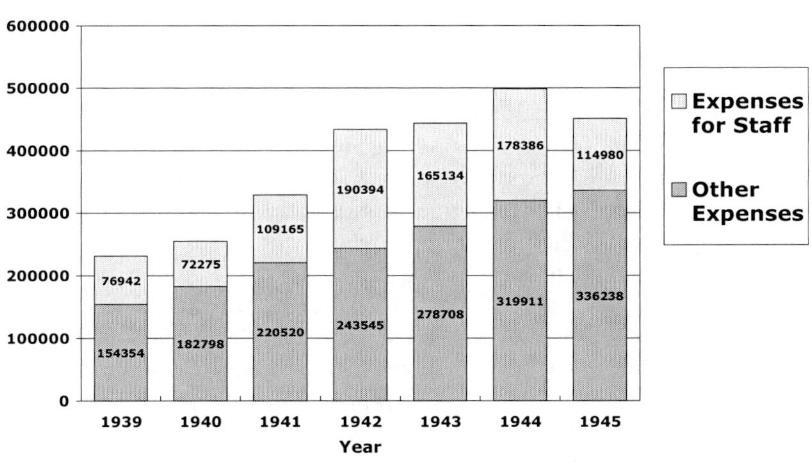

Figure 9.3 Annual budget, NLL Amsterdam 1939–1945.

Development of Staff

By accepting supervision by the German authorities in principle, not only was the directorate of the NLL able to protect its staff from being laid off, but also to expand it by almost 50 per cent by 1944. From 1939 until 1941, the staff of the NLL grew from 86 to 114 employees. Growth was only intermittent in 1942. In this exceptional year the NLL lost five engineers, three technical clerks and two workers from its general services. Since the number of instrument makers and mechanics grew by five at the same time, the overall decline of NLL staff was five employees. One might assume that this development was connected with the recruitment of forced labour by the German authorities, which was introduced in The Netherlands soon after Fritz Sauckel had been appointed *Generalbevollmächtigter für den Arbeitseinsatz* (Plenipotentiary for Labour Allocation) in March 1942.[96] Between May and December 1942, Sauckel's organisation recruited 112,600 Dutch forced labourers, who were deported to the German Reich and exploited there in the German war economy.[97] But according to the annual report of the NLL, the laboratory was not confronted with Sauckel's forced recruitments before 3 February 1943, when a small group was conscripted for work in the German Reich.[98] During a meeting of the board of the NLL in March 1943, the commercial director of the NLL, Chaillet, pleaded against an extension of the work shifts and for the employment of more staff.[99] This strategy can be seen as a response to the enforced recruitments.

At the beginning of the German recruitment policy the NLL was able, with Käufl's help, to claim the personnel and have verifications issued by the *Rüstungsinspektion Niederlande* (Armaments Inspection for The Netherlands) that protected its staff. But later in 1943, new rules were introduced that extended the age cohort for forced recruitments from Dutch employees born between 1922 and 1924 to those born between 1918 and 1921. By then, the only chance of avoiding the recruitment and deportation of the NLL staff as forced labour to the German Reich was to gain an exemption by classifying them as *Spezialkraft*.[100] In May 1943, Käufl informed Betz that all the staff of the NLL had been exempted from forced labour allocations to the Reich after he had intervened with the responsible German authorities.[101] However divergent the motivations and the interests of the AVA and the NLL were, they converged at the point of maintaining business and secure staff.

During the war, the number of engineers and academics employed at the NLL in Amsterdam showed little fluctuation, similar to the number of employees of the administration and the technical clerks. In contrast, the number of less-qualified technicians grew impressively. The number of instrument makers, mechanics and other technicians employed grew from 33 in 1940 to 68 in 1944 and thereby more than doubled. Even in the last two years of the occupation the number of employees of the NLL grew from 109 to 144, which represents an increase of about one third of the total staff employed in 1942. After liberation, the financial crisis of the NLL led to the dismissal of 15 employees, about 10 per cent of its staff in 1945.

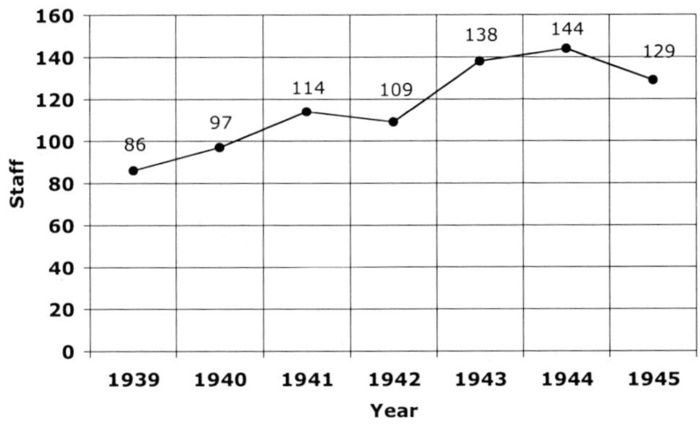

Figure 9.4 NLL staff, 1939–1945.

Development of Research Orders

Before the occupation, the income from research orders for the NLL reached a high of Hfl. 120,822 in 1939.[102] In 1940, the NLL's research income collapsed and fell almost 50 per cent to Hfl. 68,477 as a consequence of cancelled contracts and the fact that from spring to summer of 1940 the wind tunnels had to be calibrated before normal operation could begin.[103] New research projects from the AVA filled the order books the following year. Income grew to Hfl. 131,164 in 1941, when the situation was still influenced by the amount of unfinished projects from 1940.[104] The hold-up of orders was caused by the time-consuming preparatory work and the development of the required instruments.[105] In addition, the disruption of the operation of the large wind tunnel led to delays that undermined the possibility of catching up with the accumulated orders from 1940. The NLL's annual report for 1941 says that the bulk of the orders from Germany came from the AVA in Göttingen. Under the extraordinary circumstances of the occupation, the collaboration with the AVA and Käufl was seen as 'very satisfying'. In negotiations with Käufl, the NLL was allowed not only to execute measurements for the German side, but also to receive payments for detailed studies of the results. This expansion of work allowed the NLL to employ more personnel.[106] Despite the technical difficulties with the propeller of the large wind tunnel, research income rose to Hfl. 219,536 in 1942.[107] Compared to the preceding year, the income had grown more than 60 per cent. This trend continued in 1943, when the budget of research orders reached its peak with Hfl. 296,578[108] before dropping about 20 per cent to Hfl. 237,656 in 1944.[109] The figure for 1944 was already influenced by the withdrawal of the AVA from Amsterdam in September the same year. Experimental work was discontinued in the last quarter of 1944. In 1945, the NLL had difficulties obtaining enough research orders. Income fell to a meagre Hfl. 42,334, which was less than 20 per cent of the result for the preceding year.[110]

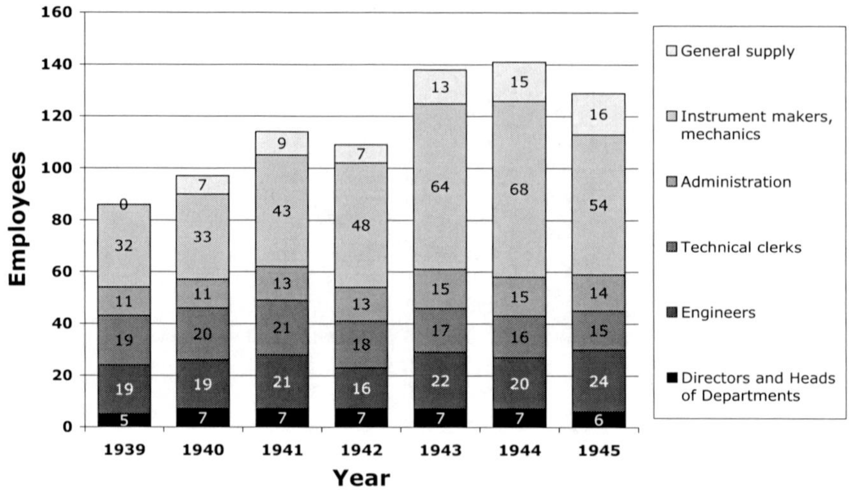

Figure 9.5 NLL staff, 1939–1945.

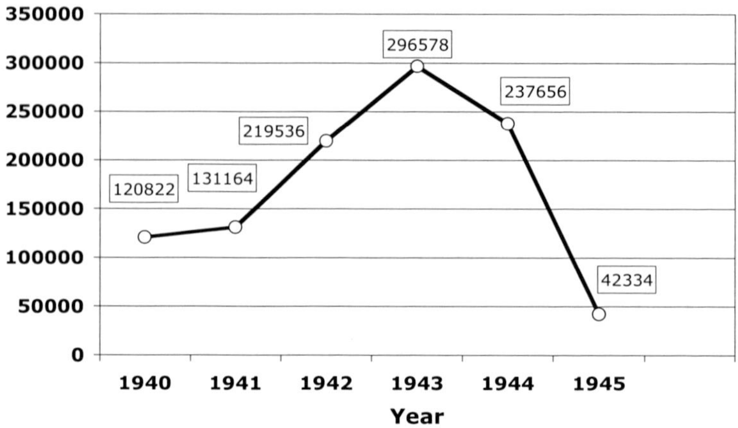

Figure 9.6 NLL income by orders, 1940–1945.

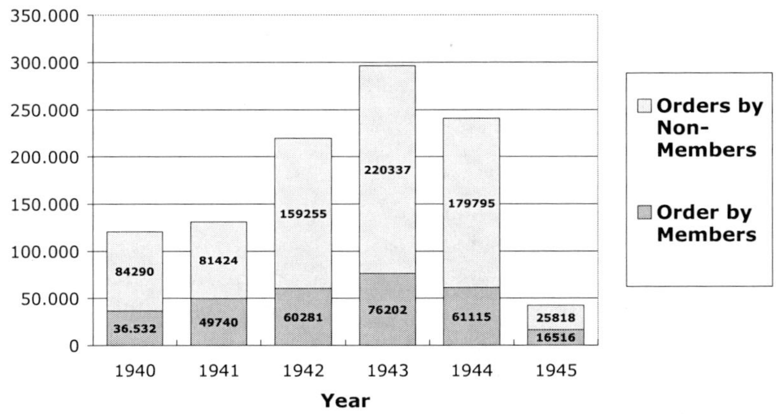

Figure 9.7 Research orders of the NLL Amsterdam, 1914–1944.

German research orders dominated the income of the NLL. Since most of the external research orders by non-members came from the AVA or German aeronautical companies, the distribution of research orders by members and non-members indicates the strong economic dependence of the NLL on Germany. The proportion of research contracts from Germany (non-members of the Stichting) reached more than 80 per cent in 1942. In March 1943, Jean Louis Chaillet stated at a board meeting of the Stichting NLL that 67 per cent of the total financial income resulted from external orders from the AVA in Göttingen, while 33 per cent came from the Dutch side.[111]

Flutter Research Ju52 and Me109

The former head of the Structure Department, Arie van der Neut, stated in his memoirs that Koning had reached an agreement with the director of the AVA, Albert Betz, that no research should be conducted in Amsterdam that would clearly be of benefit to the German war effort.[112] According to this account, Betz had issued a guideline for research contracts at the NLL that war-important research 'should not be given to the NLL since the Dutch were not to be trusted' and contracts that would have 'no direct relevance on the war' or were of a 'more general nature' should be 'labelled *kriegswichtig*'. Van der Neut admitted that no written document of the agreement was to be found in the archives, but underlined that the AVA would have 'adhered to it'.[113] The claim that the AVA and the NLL had agreed not to conduct any military research has been taken up in historiography.[114] However, it conflicts with available contemporary documents. Even though the majority of projects of the NLL in Amsterdam might not directly be identified as military research, several examples provide evidence of research projects that were clearly relevant for the German war effort.

Junkers Ju52

One research project that was relevant for the German military was the examination of the well-known Junkers Ju52 aircraft. Planned by Junkers in 1929, it was designed from the beginning as a dual-use product. The aircraft could be used for civilian and military transportation and serve as a bomber.[115] The Ju52 was used for the first time in the Spanish Civil War by the notorious Legion Condor in air raids on Madrid and Guernica (26 April 1937).[116] However, the low speed achieved in flight made the Ju52 vulnerable in aerial combat. Therefore, from May 1937 onwards, the German Air Force used the Ju52 for military transportation, but no longer as a bomber. At higher speeds, the aerodynamic design of the aircraft caused aeroelastic problems of wing flutter. Between 1937 and 1939 Junkers, together with the *Deutsche Versuchsanstalt für Luftfahrtforschung* (German Experimental Institute for Aeronautical Research) in Berlin-Adlershof, conducted a series of ground oscillation experiments and flight tests with the Ju52.[117] In July 1941, the NLL received a contract to study aeroelastic properties of the Ju52 by evaluating the results of the experiments by Junkers and the *Deutsche Versuchsanstalt für Luftfahrt* mathematically. It is remarkable that in this case the Germans carefully avoided giving any hint to the scientists of the NLL of the type of wing and aircraft company to be examined. This information was added for German readers in January 1943 on an inserted sheet attached to the NLL report by the director of the AVA Institute of Unsteady Fluid Motion, Hans-Georg Küssner. The objective of the study given to the NLL scientists was to find out under which assumptions flutter theory would be able to reproduce the results of the practical experiments by Junkers and the *Deutsche Versuchsanstalt für Luftfahrt*.[118] As mentioned by the Dutch scientists in their final report, they had received construction sketches of a wing and detailed information about its distribution of mass. Parts of the German studies originating from Junkers were also given to them. Those extracts did not give any information about the specific type of aircraft. However, the extracts included data about ground oscillation experiments and a series of wing oscillation tests, as well as a description of the changes in the construction of the aileron to improve it and avoid wing flutter oscillation.[119] In their report, completed in October 1942, Greidanus and van de Vooren stated that, from the information given, they were able to confirm that the flutter oscillation 'is mainly congruent with the anti-symmetric self-swinging and a horizontal component of this oscillation has a crucial influence on stability through an aileron coupling [*Massenkraft-Koppelung*]'.[120] The mathematical evaluation led to the conclusion that flutter theory indeed approved the results of the experimental data, qualitatively at least.[121]

Messerschmitt Me109

Unlike the Ju52 research project, the second example of a militarily relevant research project shows that, in this case, clearly no security measures were taken to keep the foreign researchers ignorant of what type of wings they were testing. In March 1941, Albert Betz, the director of the AVA, in a letter to the German Air Ministry, applied for flutter and torsion experiments at the NLL in Amsterdam with a Messerschmitt Bf109 wing. Since the Bf109 would be 'sufficiently known', Betz argued, there would be 'no concerns against the disposal of such a wing'.[122] The German Air Ministry approved the experiments the same month.[123] The Legion Condor had used the fighter plane from 1937 until 1939 in the air battle in the Spanish Civil War. Serial production of the Bf109 had begun in 1937 and reached a total of 34,000 aircraft by 1945.[124] Before the beginning of the war the aircraft had been sold to Switzerland, the Kingdom of Yugoslavia and the USSR. During the war, proliferation was restricted to the Axis powers.[125] But the Allied powers were also able to study the aircraft design in detail since a number of planes had crashed in the air battle over Great Britain between July and October 1940.[126] For this reason, special security measures to camouflage the Bf109 wing for the experiments at the NLL were unnecessary. In June 1941, three months after the German Air Ministry had approved the research programme for Amsterdam in general, Hans-Georg Küssner, head of the Institute for Unsteady Fluid Motion of the AVA, submitted more detailed information about the planned experiments to the Air Ministry.[127] The aim of the experiments was to develop a mathematical calculation to give data about the torsion stiffness of wings to calculate flutter. With its rather simple static design, the wing of the Bf109 appeared to be appropriate for such a study.[128]

In May 1941, the AVA ordered the NLL to calculate the average properties for the whole wing of the Me109 and develop a simple method for analysing systematic stiffness experiments. The experiments were important since stiffness was a crucial factor in determining the critical speed for specific wings. The idea of the study was to compare the experimental results with those achieved though calculation.[129] In June 1941, Käufl telegraphed to the AVA that enough (human) computers were available in Amsterdam. This message was probably directly connected to the planned calculations of the Bf109 that required a remarkable number of staff with certain mathematical skills.[130]

In August 1942, the report by Arie van der Neut was finished and delivered specific data about the stiffness of the Me109 wing.[131] German aeronautical companies, such as Focke-Wulff, were interested in the results and received copies of the report.[132]

As both examples show, militarily relevant research was conducted in Amsterdam, even though the Germans were careful not to reveal new secret military developments to the Dutch scientists. The experiments for the improvement of the wings of the Messerschmitt pursuit plane Me109 were possible since this military aircraft had been introduced a few years earlier and its qualities and properties were well known to the Allies.

Collaboration and Resistance as Responses towards Occupation and Repression: Conflicts with Fokker

A year after the invasion of The Netherlands by the *Wehrmacht* the German administration of the NLL was confronted with the threat of losing control of the institute to the aeronautical industry. The starting point was an incident in May 1941, when two pilots of the Dutch Fokker Company hi-jacked a Fokker G1 aircraft and flew it safely over the Channel to England.[133] When the commanding general of the *Luftgau Holland*, Hans Sigburg, was informed about this act of resistance, he proposed the arrest of family members of the pilots[134] to the Chief of the Security Police, *SS-Gruppenführer* Hanns Albin Rauter.[135] As an act of reprisal the German security police raided the KLM Pilot School in Amsterdam, confiscated documents and instruments and closed the School.[136] The Commander-in-Chief of the *Luftgaukommando* Holland ordered the arrest of all pilots in The Netherlands, radio operators and the head of Fokker plant security. Furthermore, all schools dealing with aeronautics, including the NLL in Amsterdam, were to be closed.[137] However, the implementation of this order by the Air Force failed, since *SS-Gruppenführer* Rauter insisted that the *Wehrmacht* could not interfere with security police investigations. Rauter himself was not opposed to closing the aeronautical schools, but at the same time emphasised that he was 'uninterested'.[138] Up until July 1941, the managers of the Fokker Company tried to convince General Hans Sigburg to close down the NLL and use its facilities for serial production of Bücker aircraft. To counter the threat of closing the NLL, Käufl invited General Sigburg to visit the NLL and showed him the installations of the laboratory. In late July 1941, Käufl reported to the AVA that he had been able to persuade Sigburg that 'the nature of the scientific work' of the NLL differed from that of 'a production company'.[139]

Fokker continued efforts to take over buildings or staff from the NLL. Once again, on 4 February 1943, the Technical Director, the German flight captain Dr Ing. Ernst Wilhelm Pleines, contacted the *Forschungsführung*, the department in the German Air Ministry that coordinated aeronautical research, and asked for the withdrawal of scientifically trained staff from the NLL to Fokker for work tasks fulfilled for the Messerschmitt company. Pleines disputed that the forty-two engineers working for the NLL all fulfilled war-important work.[140] After speaking with Käufl, he was convinced that Käufl would stay only temporarily at the NLL in Amsterdam. For this reason he was also not convinced that the individuals working at the NLL would proceed with more than the 'known Dutch lassitude'.[141]

The German Air Ministry remained rather unimpressed. The *Forschungsführung* agreed that work efficiency at the NLL could be improved if a permanent German head of institute could be installed. But no definite order was given. It was recommended that Pleines should be informed about the urgency of the research conducted.[142] At the same time Käufl had to respond to the increasing radicalisation of the occupation policy in The Netherlands. He was afraid that the NLL would lose its staff completely since, as a mere Dutch enterprise, it was not protected from forced labour recruitments. To secure the NLL staff, Käufl received the approval of Friedrich Seewald of the *Forschungsführung* of the German Air Ministry to transform the NLL into a German enterprise and take over staff into the AVA. Käufl considered moving from Paris to Amsterdam, since it was necessary to have a permanent representative in Amsterdam.[143]

In March 1943, after the attack by Fokker had been rejected, Käufl reported his observations on the development at the NLL in a letter to the *Forschungsführung.* He complained that 'Dutch institutions react in a hostile manner to all our suggestions', which would be 'connected with the political and military situation'. According to Käufl, 'the Dutch willingness to work' had decreased very much in the past month, such that 'one could almost speak of passive resistance'. For this reason the decision of the German Air Ministry in May 1940 to leave the Dutch character of the NLL untouched and to reject the confiscation of the institute by Fokker was now at stake.[144] Käufl demanded that the German Air Ministry change its policy towards the NLL and impose a stricter control on its staff:

> In the most recent period the situation has changed insofar as one cannot count on the willingness of the Dutch to work any more – an observation that has been made by all authorities and companies there. All the drastic measures that currently have to be taken by the security police act on the assumption that the Dutch still, and now more than ever, have to be seen as our enemy and treated as such.[145]

Käufl informed the German Air Ministry that 'over the course of time, all the German-friendly staff of the NLL have been let go for specious reasons, and the fact that the NLL has been removed from the direct control of the security police has been used to assemble a staff that has an attitude that is reactionary in principle. I do not consider it as impossible that such an island could become a danger to the state and feel obliged to let the security police know about it'.[146] To push forward the willingness of the Dutch staff to work under these circumstances, he proposed the establishment of a team of German factory security officers, and confiscation of the NLL by the AVA. Employees who were unwilling to collaborate should be deported to the German Reich as forced labourers in the context of the *Saukel-Aktion*. By regrouping the staff, Käufl judged that resistance could be broken and further interventions by the security police that might hinder continuing research could be effectively prevented.[147]

Käufl's plans were much too far-reaching for the management of the AVA in Göttingen. At the end of March 1943, the AVA decided, contrary to the suggestions of Käufl, not to confiscate the NLL. This opposition was not motivated by any concerns about brutalised repression or forced labour recruitments of the occupation policy per se. In the first place, this decision was motivated by economic reasons. The management of the AVA in Göttingen simply saw no opportunity to send the personnel necessary to control the institute in Amsterdam. Käufl received orders to influence the composition of the staff through negotiations. The management of the AVA intended to delegate a permanent representative to Amsterdam to manage research contracts from Göttingen and coordinate communication with the German occupation authorities.[148]

The strict measures Käufl had suggested to enforce control over the laboratory staff contrast with the positive appraisal of him as a defender of the autonomy of the laboratory. In the first official annual report of the NLL to be published after the war, Käufl was described as a person who had done 'his best to maintain the laboratory' and 'as much as possible to prevent external interference'.[149]

By the end of March 1943, the director of the AVA, Albert Betz, wrote a letter to the director of Fokker Amsterdam, Pleines, in which he responded to the allegations that the NLL was only conducting research work unimportant to the war. Betz underlined that the tasks fulfilled by the NLL would be 'important to war indeed' and would belong to a high level of urgency of the German war economy:

> What is disturbing and burdening to the assignment of tasks is the circumstance that the tasks must be distributed and camouflaged, so their war-important purpose cannot be realised. Thereby, naturally, for an outsider the impression might easily emerge that only unimportant things would be concerned. This will be unavoidable with regard to the crucial importance of the research work, as long as any Dutch get involved with it.[150]

In the spring of 1943, Käufl acquired an apartment in Amsterdam, which allowed him to extend his inspection visits to the NLL to several days.[151]

Resistance and Repression

In 1944, however, repression increased. On 12 April 1944, Jean Louis Chaillet was arrested by the German security service SD and imprisoned in a jail in Amsterdam.[152] Even for Käufl it was difficult to ascertain the concrete reason for the imprisonment. On 28 April 1944, he wrote to Albert Betz in a letter that, despite his efforts, the SD would only inform him that Chaillet was accused of preferential treatment of the enemy (*Feindbegünstigung*).[153] The resistance activities Chaillet had been arrested for were obviously not connected with the NLL. As the historian Louis de Jong reports, Chaillet had been a leading member of the illegal *Nationaal Comité van Verzet* (National Committee of Resistance) and was arrested because of his support of Jews who had gone into hiding as *onderduikers* (persons in hiding).[154] At a meeting with officers of the SD on 27 April 1944, Käufl had been able to obtain permission for Koning to visit Chaillet before the commercial director of the NLL was deported to a concentration camp. Käufl wrote to Betz that he saw nothing else he could do for Chaillet himself. As regards Koning, Käufl asked Betz if he could undertake something to obtain the release of Chaillet.[155] No documents have been found so far that indicate any intervention with the German authorities by Betz in favour of Chaillet. Chaillet was deported to the concentration camp in Vught and later to the Sachsenhausen concentration camp, where he arrived on 8 September 1944.[156] He survived the death march from Sachsenhausen and was liberated by Canadian troops in Lübeck in May 1945.[157] Chaillet returned to the NLL on 19 June 1945.[158]

Before he was deported to Sachsenhausen the Germans had arrested further staff members at the NLL. Adrianus Boelen, head of the Aerodynamics Department, and the secretary Ms Bé Wenting were imprisoned on 12 August 1944.[159] While Wenting was released after one week, Boelen remained in prison for six weeks. Permission to visit both was given to the NLL under the false pretence that important laboratory business matters had to be discussed.[160] In the final period of the German occupation two more employees of the NLL, the engineers J. Frits Hengeveld from the Aerodynamics Department and Izak Brinkhorst from the Structure Department, were arrested. Neither was released before the capitulation of the German troops.[161]

Apart from Chaillet, who had been arrested for political reasons as a resistance fighter, there were two Jewish employees of the NLL who were deported to concentration camps. The engineer Alfred A. Spits, who had worked for the Aircraft Department since August 1939, was deported to the Dachau concentration camp on 1 October 1941.[162] He died there of typhus in February 1945.[163] The draughtsman, Hartog Groen, who had also worked for the Aircraft Department from 1921 until 1941, was deported from the Westerbork concentration camp to the Sobibor destruction camp on 20 July 1943 and was murdered there three days later.[164] When Groen had received a letter from the Departement van Waterstaat (Ministry of Water Management) in early 1941, with a demand to determine his income after his dismissal, Chaillet protested and sent back the form with a note that Groen was still in his post at the NLL.[165] Chaillet tried to convince Käufl that the dismissal of Spitz and Groen would harm the research work done for Germany. His plan to supply home work for the two Jewish employees once they were no longer allowed to enter the NLL building was a courageous effort to maintain solidarity.[166] However, once the German occupiers began deportations of the Jewish population in The Netherlands, the resistance was unable to save their lives.

Social Life and Survival of the 'Hunger Winter' 1944/45

By the end of August 1944, Käufl was called back to Göttingen and returned to Amsterdam for the last time on 18 September 1944 for a couple of days to arrange the withdrawal of the AVA.[167] The NLL as an institution played an important role for its staff when it became more and more difficult to make ends meet in the final phase of the German occupation. The NLL functioned as a social organisation that helped its personnel to survive the shortages of food, heating, clothing and medical care during the hunger winter of 1944/45, when thousands were starved to death in Amsterdam as a consequence of the Germans plundering the food resources of The Netherlands.[168] When the delivery of food to the NLL staff canteen was discontinued in early September 1944, a committee was formed to buy foodstuffs collectively and secure supply during the hunger winter.[169]

On 19 October 1944, the electric power supply was interrupted.[170] Experimental work had to be stopped, although theoretical work continued. To further protect the staff from recruitments, prevent plundering of the instruments and secure the food supply of the staff, it was decided not to close down the NLL. The workshop was now used for the production of small ovens ('*noodkacheltjes*') and carbide lamps to help the staff come through the winter.[171] In its workshops the NLL was able to fulfil some jobs for illegal resistance organisations, mainly the maintenance and production of arms. According the first annual report of the NLL to be published after the liberation, these jobs were initially done in the NLL's workshops at weekends. Later, when transportation became too risky, the work was continued in a factory at Haarlemmermeer by an NLL instrument maker.[172] In the final period of the occupation, members of a communist group from the NLL exercised with weapons in the basement of the NLL, where a shooting stand had been installed.[173] From September 1944 onwards, important inventory and instruments were removed from the NLL to safe hiding places to protect them from plundering. The inventory was returned and reinstalled after the liberation, beginning on 14 May 1945.[174]

Concluding Remarks

To benefit immediately from the scientific resources in the summer of 1940 the German occupation authorities applied a policy that could be described as a form of indirect rule. This policy enabled German aeronautical industry and research establishments to give research orders to Dutch scientists of the NLL.

The research projects in Amsterdam gave the AVA in Göttingen and the German aeronautical industry access to additional resources for their military research in two different ways. As the flutter research project for the Ju52 and the torsion experiments with the wings of a Messerschmitt Bf109 show, direct war-relevant research was conducted. In contrast to German and Dutch claims during and after the war, secrecy measures did not prevent such projects. However, even those research projects that were not obviously connected with military research reduced the working hours of staff and wind tunnels at Göttingen and thereby represented indirect war research. By reducing the workload in Germany, additional resources were made available for military research at the AVA. For the NLL, this scientific collaboration led to a tremendous increase in staff, budget and income from external research orders.

At the same time the NLL in Amsterdam was able to survive and to continue aerodynamic research. This continuation of the NLL as an organisation came at a price. Two-thirds of the NLL's research work was done for Nazi Germany, some of it directly related to the German war effort. On the other hand, it was this institutional continuity that guaranteed resources for social and political resistance that at least improved the situation of NLL staff. With reference to the German research contracts and their war importance, employees of the NLL could be protected from recruitment to the so-called *Arbeitseinsatz* in the German Reich as forced labourers. According to the first annual report to be published after the liberation, the NLL was even in the position to employ and thereby protect scientists of other Dutch scientific institutions.[175]

Three major periods can be distinguished in the German occupation policy. The period of control through 'indirect rule' lasted from the summer of 1940 to the spring of 1943. This concept became discredited in the spring of 1943. From then on, occupation policy included enforced and brutalised repression by the German side, reacting against growing resistance that arose in the context of the German military setbacks. This second period ended in September 1944, when Josef Käufl, the *Beauftragte* of the AVA in Göttingen, returned to Germany. The management of the NLL was now able to organise social and economic support for its staff to survive the hunger winter until the liberation of The Netherlands in May 1945.

Acknowledgements

Research for this study has been made possible by the Deutsche Forschungsgemeinschaft (Schwerpunktprogramm 1143: Wissenschaft, Politik und Gesellschaft. Deutschland im internationalen Zusammenhang im späten 19. und im 20. Jahrhundert: Personen, Institutionen, Diskurse). The author would like to express his gratitude to DirkJan Rozema, Kees Bakker and Jan Hoefnagels (Stichting Historisch Museum Nationaal Lucht- en Ruimtevaartlaboratorium, Amsterdam), Dr Sandra Sacher (Nationaal Archief, Den Haag), Jessika Wichner and Heinz Fütterer (Archives of the Deutsche Luft und Raumfahrtzentrum, Göttingen), Dr Wilhelm Füßl and Dr Eva A. Mayring (Archives of the Deutsche Museum, Munich) for their kind support to find relevant sources for this study.

Notes

1 L. Budraß, *Flugzeugindustrie und Luftrüstung in Deutschland 1918–1945* (Düsseldorf: Droste, 1998), 293–375; E. L. Homze, *Arming the Luftwaffe: The Reich Air Ministry and the German Aircraft Industry, 1919–1939* (Lincoln, Neb.: University of Nebraska Press, 1976), 209–16.

2 C. Tollmien, 'Luftfahrtforschung: Die Aerodynamische Versuchsanstalt in Göttingen', M. Tschimer (ed.), *Wissenschaft im Krieg – Krieg in der Wissenschaft. Ein Symposium an der Philipps-Universität Marburg. 50 Jahre nach Beginn des 2. Weltkrieges* (Marburg,1990), 64–79; C. Tollmien, ‚Das KWI für Strömungsforschung verbunden mit der aerodynamischen Versuchsanstalt', H. Becker, H.-J. Dahms and C. Wegeler (eds), *Die Universität Göttingen unter dem Nationalsozialismus. Das verdrängte Kapitel ihrer 250-jährigen Geschichte* (München: Saur, 1998), 684–708; H. Trischler, 'Die Luftfahrtforschung im Dritten Reich. Organisation, Steuerung und Effizienz im Zeichen von Aufrüstung und Krieg', H. Boog (ed.) *Luftkrieg und Luftkriegführung im Zweiten Weltkrieg. Ein internationaler Vergleich* (Herford: Mittler, 1992), 205–33; H. Trischler, *Luft- und Raumfahrtforschung in Deutschland 1900–1970. Politische Geschichte einer Wissenschaft* (Frankfurt am Main: Campus, 1992), 199–207; M. Epple, 'Rechnen, Messen, Führen. Kriegsforschung am Kaiser-Wilhelm-Institut für Strömungsforschung 1937–1945', H. Maier (ed.), *Rüstungsforschung im Nationalsozialismus* (Göttingen: Wallstein-Verlag, 2002), 305–56.

3 J. C. Rotta, *Die Aerodynamische Versuchsanstalt in Göttingen, ein Werk Ludwig Prandtls. Ihre Geschichte von den Anfängen bis 1925* (Göttingen: Vandenhoeck & Ruprecht, 1990); Trischler 1992, 133.

4 Trischler 1992, 201.

5 Figures for the staff (dating from 31 March 1938) vary in the 1937/38 annual report, cf. Archives of the Max Planck Society (AMPS), I. Abt., Rep. 2, Nr. 180, 2 and the 1938/39 annual report, cf. AMPS, I. Abt., Rep. 2, Nr. 1812. The 1937/38 annual report gives a figure of 544 and the report for 1938/39 a figure of 534 employees. See also: Wfm. (AVA Göttingen), Personalbestand, 21 January 1941, Archives of the Deutsche Luft- und Raumfahrtzentrum (DLR) Göttingen (in the following referred to as DLR Archives), GOAR 3221, fot. 34.

6 Wfm. (AVA Göttingen), Personalbestand, 21 January 1941, DLR Archives, GOAR 3221, fot. 34. Figures for the number of staff (status 31 March 1939 and 31 March 1940) were mixed in the annual report for 1939/40, AMPS, I. Abt., Rep. 2, Nr. 182, 5. The figures given by Trischler differ slightly. He speaks of a staff of 80 employees in 1933, reaching 450 in 1936 and more than 700 by the end of the war. Cf. H. Trischler, 'Aeronautical Research under National Socialism: Big Science or Small Science?', M. Szöllösi-Janze (ed.) *Science in the Third Reich* (Oxford: Berg, 2001), 79–110, 89 and Trischler 1992, 202.

7 Trischler 1992, 196.

8 AVA Göttingen, Organogramm Stand 1.11.1941, DLR Archives, GOAR 3221, fot. 4.

9 See the collection of documents concerning the AVA-Außenstelle Prag, DLR Archives, GOAR 788, fot. 111–30.

10 Die Einteilung der AVA in Institute seit 1939, DLR Archives, GOAR 3221, fot. 15.

11 For the establishment of the outpost at Finse in the autumn of 1940, cf. Ritz, Aktenvermerk, Betr.: Kälteforschung innerhalb der AVA, 29.6.1942, DLR Archives, GOAR 788, fot. 107–9 and correspondence, DLR Archives, GOAR 788, fot. 111–30.

12 CV of Michael Strscheletzky, born 12 October 1904 in Simbirsk, 7 August 1944, DLR Archives, GOAR 3186, fot. 2 and GOAR 2730.

13 K. Schlör von Westhoffen-Drimstein: ‚Der Deutsche Motor-Schlitten. Geschichte der Aussenstelle Riga der Aerodynamischen Versuchsanstalt Göttingen', *DLR IB*, 20002-92A01 (1992), 1–98.

14 Albert Betz, Zeugnis für Herrn Dipl.-Ing. Josef Käufl, 12 November 1945, DLR Archives, GOAR, Personal file of Josef Käufl, born 5 January 1907, fot. 17–18; Questionnaire Josef Käufl, 5 February 1935, ibid., fot. 38–9.

15 Notulen van de 8de vergadering van het Bestuur van het bestuur van de Stichting ‚Nationaal Luchvaartlaboratorium', gehouden te ‚s-Gravenhage, 12 July 1940, 2, NLL Vergaderingen Bestuur, Vol. 1940–1945, Stichting Historisch Museum NLR (Amsterdam).

16 Message to Mr Engelbrecht (management of the AVA), 22 December 1944, DLR Archives, GOAR personal file of Josef Käufl, fot. 2.

17 Baeumker (Reichsminister der Luftfahrt, LC1, Az 55a 0010/Nr. 0070 40/40 g (IA)) to Luftgaukommando Holland (Amsterdam), copy, 25 May 1940, Nationaal Archief (Den Haag), Directoraat-Generaal van de Rijksluchtvaartdienst, toegang 5.016.5240, bestanddeel 7691.

18 Käufl to AVA Göttingen über LC 1, Bericht über das Nationale Luftfahrt-Laboratorium Amsterdam (LC 1/107326/41g), 6 June 1940, DLR Archives, GOAR 2729, fot. 567.

19 Koning was appointed 1 August 1940. Cf. J. A. Van der Bliek, *75 Years of Aerospace Research in The Netherlands. Nationaal Aerospace Laboratory, NLR 1919–1994* (Amsterdam: Joh. Enschedé, 1994), 31.

20 P. Longerich, *Politik der Vernichtung. Eine Gesamtdarstellung der nationalsozialistischen Judenverfolgung* (München, Zürich: Piper, 1998), 499; and B. Moore, *Victims and Survivors. The Nazi Persecution of the Jews in the Netherlands, 1940–1945* (London: Arnold, 1997), 54–8.

21 Van der Bliek 1994, 31 and 230–4. For the beginning of the deportation of the Dutch Jews see: Longerich 1998, 499 and Moore 1997, 66.

22 Käufl, Bericht über das Nationale Luftfahrt – Laboratorium Amsterdam, 6 June 1940, DLR Archives, GOAR 2729, fot. 568 and Verslag over het jaar 1940 van de Stichting 'Nationaal Luchtvaart-Laboratorium' (NLL), Amsterdam 1941, 11, Stichting Historisch Museum NLR (Amsterdam).
23 Verslag over het jaar 1940 van de Stichting 'Nationaal Luchtvaart-Laboratorium' (NLL), Amsterdam 1941, 11, Stichting Historisch Museum NLR (Amsterdam).
24 For the architectural history of the NLL building and its architect, Hugh Maaskant, cf. chapter 3 in: M. Provoost, *Hugh Maaskant. Archivtect van de vorruitgang* (Rotterdam: Uit. 010, 2003).
25 Käufl to AVA Göttingen via Reichsluftfahrtministerium (Reichsluftfahrtministerium - LC 1), Bericht über das Nationale Luftfahrt-Laboratorium Amsterdam (LC 1/107326/41g), 6 June 1940, DLR Archives, GOAR 2729, fot. 567–9.
26 Ibid., fot. 568.
27 Van der Bliek 1994, 29, 52. The author does not give a source for this assumption. It therefore seems more likely that the existence of the laboratory aircraft was well known by the Germans, as indicated in the German document cited above. The existence of the aircraft is also mentioned in the published report of the NLL by T. van Oosterom: 'Het Nationaal Luchvaartlaboratorium te Amsterdam', *Marineblad*, 57 (1942), 19.
28 AVA to NLL, 13 June 1941, DLR Archives, GOAR 363, fot. 191 and NLL to AVA, Betr.: Belegung der Windkanäle, 4 April 1941, DLR Archives, GOAR 363, fot. 192 and Verslag over het jaar 1940 van de Stichting 'Nationaal Luchtvaart-Laboratorium' (NLL), Amsterdam 1941, 44, Stichting Historisch Museum NLR (Amsterdam).
29 A. de Lathouder: 'De groote windtunnel van het Nationaal Luchvaartlaboratorium (NLL rapport A.795)', *Verslagen en verhandelingen - Nationaal Luchtvaartlaboratorium*, Teil XI (1942), A1 and van Oosterom 1942, 42.
30 For the old RSL wind tunnel No. 1 see: E.B. Wolff and C. Koning: 'De technische beteekenis van groote windtunnels (NLL rapport A.544)', *Verslagen en verhandelingen - Nationaal Luchtvaartlaboratorium*, VIII (1939), 2–14. For wind tunnels Nos. 2–4 see: de Lathouder 1942.
31 Bericht über die Besichtigung der strömungstechnischen Forschungseinrichtungen in Amsterdam, Wageningen, Brüssel, Lille und Paris, in der Zeit vom 18. bis 27. Juli 1940, undated, DLR Archives, GOAR 2729, fot. 542–6.
32 Betz to Reichsluftfahrtministerium (LC 1), 2 August 1940, DLR Archives, GOAR 2729, fot. 548.
33 Appendix to the letter from Betz to Reichsluftfahrtministerium (LC 1), 2 August 1940, DLR Archives, GOAR 2729, fot. 548–52.
34 Ibid.
35 Baeumker to AVA, Betr.: Forschungsprogramm, 6 August 1940, DLR Archives, GOAR 2729, fot. 547.
36 Baeumker (Reichsluftfahrtministerium - Technisches Amt - LC 1 II E) to AVA, Betr.: Auftragsvergebung an das Nationale Luftfahrtlaboratorium Amsterdam (NLL), 2 September 1940, DLR Archives, GOAR 2729, fot. 431.
37 Protokoll einer Besprechung bei Regierungspräsident Dr. Stüler, Beauftragter des Reichskommissars für die besetzten niederländischen Gebiete Seyß-Inquart, unsigned, Göttingen, 20 September 1940, DLR Archives, GOAR 2729, fot. 523–4; and Stüler to Generalsekretär Dr Frederiks (Ministerium des Inneren), Betr. NLL Amsterdam (carbon copy), 21 January 1941, Nationaal Archief (Den Haag), Directoraat-Generaal van de Rijksluchtvaartdienst, toegang 5.016.5240, inventarisnummer 7691.
38 Baeumker (Reichsluftfahrtministerium) to Reichskommissar für die besetzten niederländischen Gebiete - Generalkommissar für Verwaltung und Justiz, 19 December 1940, DLR Archives, GOAR 2729, fot. 432–3.
39 vR (NLL): Notulen van de negende vergadering van het Bestuur van het Stichting 'Nationaal Luchvaartlaboratorium', te houden op 25 April 1941 te Amsterdam, 7–8,

Stichting Historisch Museum NLR (Amsterdam), NLL Vergaderingen Bestuur, Vol. 1940–1945.

40 Verslag over het jaar 1940 van de Stichting 'Nationaal Luchtvaart-Laboratorium' (NLL), Amsterdam 1941, 58, Stichting Historisch Museum NLR (Amsterdam).

41 Keuringsprogramma NLL voor Verf- en Laklagen (15 September 1940), 3 October 1940, NLL report M.963; Overzicht over den inhoud van het artikel 'Härteverfahren', 8 May 1941, NLL report M. 974; Zug-Druck-Dauerversuche an Bolzen-Augenstäben. Teil I, 30 September 1941, NLL report M.984, and Statische Versuche an Bolzenverbindungen, 14 August 1944, NLL report M.1026, Stichting Historisch Museum NLR (Amsterdam).

42 Verslag over het jaar 1940 van de Stichting 'Nationaal Luchtvaart-Laboratorium' (NLL), Amsterdam 1941, 55–9.

43 Verkort verslag over het jaar 1941 van de Stichting 'Nationaal Luchtvaart-Laboratorium' (NLL), Amsterdam 1942, 15, Stichting Historisch Museum NLR (Amsterdam).

44 Verkort Verslag over het jaar 1942 van de Stichting 'Nationaal Luchtvaart-Laboratorium' (NLL), Amsterdam 1943, 15, Stichting Historisch Museum NLR (Amsterdam).

45 Verslag over het jaar 1940 van de Stichting 'Nationaal Luchtvaart-Laboratorium' (NLL), Amsterdam 1941, 53, Stichting Historisch Museum NLR (Amsterdam).

46 A. van der Neut, 'Die elastische Stabilität der längs- und querversteiften Zylinderschale bei axialer Druckbelastung Teil II', 1 October 1942, NLL report S. 255 and A. van der Neut, 'Die elastische Stabilität der längs- und querversteiften Zylinderschale bei axialer Druckbelastung Teil I', 2 March 1942, NLL report S.256, Stichting Historisch Museum NLR. See also A. van der Neut, *Het Nationaal Luchtvaartlaboratorium in oorlogstijd*, 1983, 9, Manuscript, Stichting Historisch Museum NLR and Van der Bliek 1994, 30–1.

47 I. Binkhorst, 'Versuche mit einem Aussenflügel einfacher Konstruktion zur Bestimmung der Verdrehsteifigkeit in ihrer Abhängigkeit vom Verdrehmoment', 31 August 1942, NLL report S.259; I. Binkhorst, 'Versuche mit einem Aussenflügel einfacher Konstruktion zur Bestimmung der Verdrehsteifigkeit in ihrer Abhängigkeit von der Verdrehungs- und Biegungsbelastung. Versuche mit dem ersten Flügel', 1 November 1942, NLL report S.267; I. Binkhorst, 'Versuche mit einem Aussenflügel einfacher Konstruktion zur Bestimmung der Verdrehsteifigkeit in ihrer Abhängigkeit vom Verdrehmoment Versuche mit dem zweiten Flügel', 1 December 1942, NLL report S.272; I. Binkhorst, 'Versuche mit einem Aussenflügel einfacher Konstruktion zur Bestimmung der Verdrehsteifigkeit in ihrer Abhängigkeit von der Verdrehungs- und Biegungsbelastung. Versuche mit dem zweiten Flügel', 1 March 1943, NLL report S.275, Stichting Historisch Museum NLR (Amsterdam).

48 A. van der Neut, 'Vergleich zwischen den Verdrehsteifigkeiten eines Flügels nach einigen Zugfeldtheorien und nach statischen Versuchen', 1 July 1944, NLL report S.290 and A. van der Neut, 'Die Schwingungsverdrehsteifigkeit eines Flügels in ihrer Abhängigkeit vom Vorlastverdrehmoment. I. Entwurf der Versuchsvorrichtung', 19 April 1944, NLL report S.291, Stichting Historisch Museum NLR (Amsterdam).

49 Verslag over het jaar 1940 van de Stichting 'Nationaal Luchtvaart-Laboratorium' (NLL), Amsterdam 1941, 23.

50 Verslag over het jaar 1940 van de Stichting 'Nationaal Luchtvaart-Laboratorium' (NLL), Amsterdam 1941, 24; and Numerische Auswertung der Gleichungen einer Flatterschwingung, 21 November 1940, NLL report V.1242b and Ergänzung zur numerischen Auswertung der Gleichungen einer Flatterschwingung, 13 January 1941, NLL report 1242a, Stichting Historisch Museum NLR (Amsterdam).

51 J. H. Greidanus, 'Die Vorbereitung von Versuchen im Windkanal zur Bestimmung der instationären Luftkräfte auf schwingende Flügel', 26 May 1941, NLL report V.1260, Stichting Historisch Museum NLR (Amsterdam).

52 T. van Oosterom, 'Vorentwürfe eines kombinierten gyroskopischen Längs- und Querneigungsmessers mit photographischer Registrierung', 15 June 1942, NLL report V.1283, Stichting Historisch Museum NLR (Amsterdam).

53 J. H. Greidanus and A. I. van de Vooren, 'Berechnung der Flatterschwingungen, beschrieben in den Versuchsberichten Nr.934/13 und 933/29 der Aerodynamischen Versuchsanstalt Göttingen', 17 October 1943; J. H. Greidanus; A. I. van de Vooren, 'Erste Ergänzung zum Bericht V.1284. Berechnung der Flatterschwingungen, beschrieben in den Versuchsberichten Nr. 934/13 und 933/29 der Aerodynamischen Versuchsanstalt, Göttingen', 1 April 1943; and J. H. Greidanus and A. I. van de Vooren, 'Zweite Ergänzung zum Bericht V.1284. Berechnung der Flatterschwingungen, beschrieben in den Versuchsberichten Nr.934/13 und 933/29 der Aerodynamischen Versuchsanstalt Göttingen', 15 May 1943, Stichting Historisch Museum NLR (Amsterdam).

54 T. van Oosterom, 'WW: Entwicklung eines registrierenden Variometers mit sehr kleiner Anzeigeverzögerung. Teil I: Voruntersuchung und Vorentwurf eines Stauscheibengeräts', 6 October 1943, NLL report V.1316a; and T. van Oosterom, 'Kurzbericht einer vorläufigen Prüfung eines Hitzdraht-Variometersystems unter Verwendung von Platindrähten', 21 February 1944, NLL report V.1342, Stichting Historisch Museum NLR (Amsterdam).

55 T. van Oosterom, 'Prüfung von zwei Typen mechanischer Übertragungen in Hinschicht auf die Verwendung im gyroskopischen Längs- und Querneigungsschreiber mit photographischer Registrierung', 5 August 1943, NLL report V.1326, Stichting Historisch Museum NLR (Amsterdam).

56 The results of these measurements were published later in 1942 and 1943: A. de Lathouder and S. I. Wiselius, 'Beproeving van windtunnel 3 van het Nationaal Luchvaartlaboratorium (NLL rapport A.796)', *Verslagen en verhandelingen - Nationaal Luchtvaartlaboratorium*, Teil XI (1942), A11–A18; and A. de Lathouder and S. I. Wiselius, 'Beproeving van het windtunnel 2 van het Nationaal Luchvaartlaboratorium. (NLL rapport V.832)', *Verslagen en verhandelingen – Nationaal Luchtvaartlaboratorium*, Teil XII (1943), A.5–A.16, Stichting Historisch Museum NLR (Amsterdam).

57 Verslag over het jaar 1940 van de Stichting 'Nationaal Luchtvaart-Laboratorium' (NLL), Amsterdam 1941, 36, Stichting Historisch Museum NLR (Amsterdam).

58 Verkort Verslag over het jaar 1941 van de Stichting 'Nationaal Luchtvaart-Laboratorium' (NLL), Amsterdam 1942, 7, Stichting Historisch Museum NLR (Amsterdam).

59 Verkort Verslag over het jaar 1941 van de Stichting 'Nationaal Luchtvaart-Laboratorium' (NLL), Amsterdam 1942, 13, Stichting Historisch Museum NLR (Amsterdam).

60 Verkort Verslag over het jaar 1943 van de Stichting 'Nationaal Luchtvaart-Laboratorium' (NLL), Amsterdam 1944, 10, Stichting Historisch Museum NLR (Amsterdam).

61 Dreikomponentenmessungen an einem Modell der Radschneekufe 8-3909, 52 Fg K II. für "Maatschappij van Berkel's Patent N.V." Gliederung, NLL report A.788, 22 April 1941, Stichting Historisch Museum NLR (Amsterdam).

62 JAW, Dreikomponentenmessungen an zwei, leicht geänderten Modellen der Schneekufe 8-3906-140 für Maatschappij Van Berkel's Patent N.V., NLL report A.876, 6 October 1942 and JAW, Untersuchung des Einflusses des Dicke- und Seitenverhältnis einer Schneekufe auf den Verlauf der Kippmomentenkurve, NLL report A.878, 19 October 1942, Stichting Historisch Museum NLR (Amsterdam).

63 J. F. Hengeveld/HB, 'Messungen an einem Flügel mit Kühler und Kühlerkanalen verschiedener Form für die Aerodynamische Versuchsanstalt zu Göttingen', NLL report A.825, 1 September 1941, Stichting Historisch Museum NLR (Amsterdam). The fact that this contract (IA 10004) was given by Messerschmitt is mentioned in:

Berichte des NLL über AVA-eigene Aufträge und über die AVA Göttingen erteilte Aufträge fremder Firmen. 31 December 1942, DLR Archives, GOAR 2729, fot. 454.

64 A. de Lathouder and S. I. Wiselius, 'Prüfung des Windtunnels 3 des NLL', NLL report A.796, February 1942, Archiv Deutsches Museum (Munich), ZWB NLL A.796.

65 C. Koning, 'Die Kanalkorrektur für einen Propeller in einem 'gemischten' Windkanal. Auftrag 232.072d, NLL report A.589', 18 May 1942, Stichting Historisch Museum NLR (Amsterdam).

66 J. F. Hengeveld, 'Bestimmung des Wandeinflusses für drei verschiedene Formen der Meßstrecken des NLL-Windkanals 4 für die AVA, NLL report A.875', 4 August 1943, Archiv Deutsches Museum (Munich), ZWB NLL A.875.

67 Untersuchung des Modells Ar 233 im großen Windkanal des N.L.L. in Amsterdam für die Aerodynamische Versuchstalt in Göttingen Teil I Messungen ohne laufende Schraube, NLL report A.914, 15 May 1943 and Untersuchung der Schrauben des Arado-Modells Ar.233 für die Aerodynamische Versuchstalt zu Göttingen, NLL report A.915, 25 May 1943, Stichting Historisch Museum NLR (Amsterdam).

68 Untersuchung des Modells Ar. 233 im großen Windkanal des N.L.L.'s in Amsterdam für die Aerodynamische Versuchsanstalt in Göttingen Teil III Vorläufige Messungen mit laufenden Schrauben, NLL report A.926, 24 July 1943, Stichting Historisch Museum NLR (Amsterdam).

69 H. J. Nowarra and K. Kens, *Die deutsche Luftrüstung 1933–1945* (München: J.F. Lehmans Verlag, 1977), 62; and H.-J. Becker, *Wasserflugzeuge – Flugboote, Amphibien, Schwimmerflugzeuge. Entwicklungsgeschichte der deutschen Flugboote, Schwimmerflugzeuge, Amphibien, Bodeneffektgeräte sowie Bordflugzeuge* (Bonn: Bernard & Graefe, 1994), 205–6.

70 Versuche zur Feststellung der auftriebssteigernden Wirkung einer Wölbungsklappe bzw. einer Spreizklappe an einem NACA 23015-Profil Ier Teil: Kräftemessungen, NLL report A.786, 7 April 1943; and J. F. Hengeveld, 'Versuche zur Feststellung der auftriebssteigernden Wirkung einer Wölbungsklappe bzw. einer Spreizklappe an einem NACA 23015-Profil Teil II: Druckmessungen, NLL report A.897', 4 August 1943, Stichting Historisch Museum NLR (Amsterdam).

71 JS, Messungen an einem Rechteckflügel mit Bremschraube in der offenen Meßstrecke des großen N.L.L. -Windkanals Teil I Allgemein. Dreikomponentenmessungen, NLL report A.909, 28 October 1943; and Messungen an einem Rechteckflügel mit Bremsschraube in den offenen Meßstrecke des großen N.L.L.-Windkanals, NLL report A.938, 17 November 1943, Stichting Historisch Museum NLR (Amsterdam).

72 Messungen an acht Pfeilflügeln für die Aerodynamische Versuchsanstalt Göttingen e.V., NLL report A.927, 2 December 1943, Stichting Historisch Museum NLR (Amsterdam).

73 J. F. Hengeveld, 'Messungen an einem Flügelmodell mit Landeklappe und zwei Unterbrecherklappen. Auftrag A 451.053 (ZWB NLL A.941 = AVA 44/F/07), 10 May 1944, Stichting Historisch Museum NLR (Amsterdam).

74 Since Albert Betz and Ludwig Prandtl were absent, Koning was shown around by Hans-Georg Küssner, the head of the AVA Institute of Unsteady Fluid Motion. See: Nationaal Luchtvaartlaboratorium: Rapport R.B. 271: Eenige besprekingen in Duitschland in Augustus 1938, 4 January 1939, 1. For the decision of the board in 1938 to build a new laboratory see: van Oosterom 1942, 38.

75 For the biography of Jakob Ackeret see: A. Betz, 'Ackeret in Göttingen', *Zeitschrift für Angewandte Mathematik und Physik*, 9b (1958), 34–6; N. Rott, 'J. Ackeret und die Geschichte der Machschen Zahl', *Schweizer Ingenieur und Architekt*, 21 (1983), 591–4; and K. Nickelsen, A. Hool and G. Graßhoff, *Flugzeuge für die Welt und eine Stiftung für Bern* (Basel: Birkhäuser, 2004), 165–76.

76 de Lathouder 1942, A1–A2, Stichting Historisch Museum NLR (Amsterdam).

77 de Lathouder 1942, A6, A8.

78 For biographical aspects and Burgers' left-wing political engagement see: J. M. Burgers: 'Some Memoirs of Early Work in Fluid Mechanics at the Technical University of Delft', *Annual Review of Fluid Mechanics*, 7 (1975), 1–12; G. Alberts: 'On Connecting Socialism and Mathematics: Dirk Struik, Jan Burgers, and Jan Tinbergen', *Historica Mathematica*, 21 (1994), 280–305; and G. Somsen, ‚Waardevolle wetenschap. Bespiegelingen over natuurwetenschap, moraal en samenleving in de aanloop naar de doorbraak-beweging', G. Somsen (ed.), *De Doorbraak van de Experts. Wetenschap en Maatschappelijke Vernieuwing Rond 1945* (Rotterdam: Erasmus Publishing, 2001), 207–24.

79 According to the annual report of the NLL, the damage occured on 10 September 1941. Cf. SL/dM: Notulen van de tiende vergadering van het bestuur van de Stichting 'Nationaal Luchtvaartlaboratorium' gehouden op 21 April 1942 te 's-Gravenhage, 11, Stichting Historisch Museum NLR (Amsterdam), NLL Vergaderingen Bestuur, Vol. 1940–1945. The same date is given by Koning to Rüstungsinspektion Niederlande, 18 November 1942, Nationaal Archief (Den Haag), Centraal Archief Bijzondere Rechtspleging, 2.09.09, inventarisnummer 107348. dossier 68212 Procurateur Fiscal Amsterdam: Nationaal Luchtvaartlaboratorium. In contrast, the minutes of the board meeting of the Stichting NLL gives 16 September 1941 as the date when the accident occurred. Cf. Verkort Verslag over het jaar 1941 van de Stichting 'Nationaal Luchtvaart-Laboratorium' (NLL), Amsterdam 1942, 13, Stichting Historisch Museum NLR (Amsterdam).

80 Verkort Verslag over het jaar 1941 van de Stichting 'Nationaal Luchtvaart-Laboratorium' (NLL), Amsterdam 1942, 7, Stichting Historisch Museum NLR (Amsterdam).

81 Verkort Verslag over het jaar 1941 van de Stichting 'Nationaal Luchtvaart-Laboratorium' (NLL), Amsterdam 1942, 13, Stichting Historisch Museum NLR (Amsterdam).

82 SL/dM: Notulen van de tiende vergadering van het bestuur van de Stichting 'Nationaal Luchtvaartlaboratorium' gehouden op 21 April 1942 te 's-Gravenhage, 11, Stichting Historisch Museum NLR (Amsterdam), NLL Vergaderingen Bestuur, Vol. 1940–1945.

83 Koning (NLL) to AVA, 25 March 1942, DLR Archives, GOAR 2729, fot. 474 and Seiferth (AVA), Notice: Ferngespräch mit Prof. Koning, NLL Amsterdam, 27 March 1942, DLR Archives, GOAR 363, fot. 169.

84 SL/dM: Notulen van de tiende vergadering van het bestuur van de Stichting 'Nationaal Luchtvaartlaboratorium' gehouden op 21 April 1942 te 's-Gravenhage, p. 11, Stichting Historisch Museum NLR (Amsterdam), NLL Vergaderingen Bestuur, Vol. 1940–1945.

85 Verkort Verslag over het jaar 1942 van de Stichting 'Nationaal Luchtvaart-Laboratorium' (NLL), Amsterdam 1943, 8. See also: J. H. Greidanus and A. van der Neut, ‚Trillingen van de schroef van windtunnel 3 Trillingsmetingen aan de schroef en de door berekening gecontroleerde interpretatie van hun uitcomsten', 3 August 1942, NLL Report A.872, Stichting Historisch Museum NLR (Amsterdam).

86 Koning (NLL) to AVA, Betr.: Auftrag Junkers Werke, 20 October 1942, DLR Archives, GOAR 2729, fot. 483

87 Lathouder mentiones the case of measurements of airplane skis for the Russian front. Cf. Notes by Adriaan de Lathouder: Bespreking Jan van der Bliek, DirkJan Rozema and Adriaan de Lathouder on 17 January 2003, 3–4, Stichting Historisch Museum NLR (Amsterdam).

88 G. Hirschfeld, *Nazi Rule and Dutch Collaboration* (Oxford u.a.: Berg, 1988), 182.

89 Käufl to AVA Göttingen über LC 1, Bericht über das Nationale Luftfahrt-Laboratorium Amsterdam (LC 1/107326/41g), 6 June 1940, DLR Archives, GOAR 2729, fot. 567.

90 Nationaal Luchtvaartlaboratorium. Verlies - en Winstrekening 1939, Anlage zu LC 1/107326/40g, DLR Archives, GOAR 2729, fot. 587 and Verslag over het jaar 1939

van de Stichting 'Nationaal Luchtvaart-Laboratorium' (NLL), Amsterdam 1940, 16, Stichting Historisch Museum NLR (Amsterdam).

91 Begrooting van de Stichting 'Nationaal Luchtvaartlaboratorium voor het jaar 1940, DLR Archives, GOAR 2729, fot. 606–7.

92 Ontwerp-begrooting voor het jaar 1943 van de Stichting Nationaal Luchtvaartlaboratorium, 27 February 1942, 1–2, Stichting Historisch Museum NLR (Amsterdam), NLL Vergaderingen Bestuur, Vol. 1940–1945.

93 Ontwerp-begrooting voor het jaar 1944. 1. maart 1944, 19 March 1944, Stichting Historisch Museum NLR (Amsterdam), NLL Vergaderingen Bestuur, Vol. 1940–1945.

94 RV (NLL): Ontwerp-begrooting voor 1945, 1 March 1944, Stichting Historisch Museum NLR (Amsterdam), NLL Vergaderingen Bestuur, Vol. 1940–1945.

95 RvW/CDD (Stichting Nationaal Luchtvaartlaboratorium), Voorloopige resultatenrekening 1944, Stichting Historisch Museum NLR (Amsterdam), NLL Vergaderingen Bestuur, Vol. 1940–1945.

96 For the appointment of Sauckel on 21 March 1942 see: D. Eichholtz, *Geschichte der deutschen Kriegswirtschaft 1939–1945. Bd. II* (1941–1943) (Berlin (DDR):Akademie-Verlag, 1985), 74–9; and U. Herbert, *Fremdarbeiter. Politik und Praxis des ‚Ausländereinsatzes' in der Kriegswirtschaft des Dritten Reiches* (Berlin: Verlag J. H. W. Dietz Nachfolger, 1986), 149–54.

97 G. Hirschfeld, 'Die niederländischen Behörden und der Reichseinsatz', U. Herbert (ed.), *Europa und der ‚Reichseinsatz'. Ausländische Zivilarbeiter, Kriegsgefangene und KZ-Häftlinge in Deutschland 1938–1945* (Essen: Klartext-Verlag, 1991), 179.

98 Verkort Verslag over het jaar 1943 van de Stichting 'Nationaal Luchtvaart-Laboratorium' (NLL), Amsterdam 1944, 4–5, Stichting Historisch Museum NLR (Amsterdam).

99 SL/dM: Notulen van de elfde vergadering van het bestuur van de Stichting 'Nationaal Luchtvaartlaboratorium' gehouden op 26 March 1943 te Amsterdam, 10, Stichting Historisch Museum NLR (Amsterdam), NLL Vergaderingen Bestuur, Vol. 1940–1945.

100 For the recruitment of the age cohorts born between 1918 and 1924 see: B. A. Sijes, *De Arbeidsinzet. De Gewongen Arbeid van Nederlanders in Duitsland, 1940–1945* (‚s-Gravenhage: Martinus Nijhoff, 1966), 282–92. For the definition of skilled or specially trained workers as »Spezialkräfte«, see: 284.

101 Käufl (Amsterdam) an Betz (AVA), 13.5.1943, DLR Archives, GOAR 1033, fot. 52.

102 Nationaal Luchtvaartlaboratorium. Verlies - en 1939, Anlage zu LC 1/107326/40g, DLR Archives, GOAR 2729, fot. 587.

103 Verslag over het jaar 1940 van de Stichting 'Nationaal Luchtvaart-Laboratorium' (NLL), Amsterdam 1941, 21; and Chaillet, Ontwerpbegrooting van de Stichting Nationaal Luchtvaartlaboratorium voor het jaar 1942, 15 February 1941, Stichting Historisch Museum NLR (Amsterdam), gives a slightly higher figure for 1940 of approx. hfl. 69,000.

104 Stichting NLL: Exploitatierekening over het boekjaar 1.1.–31.12.1941, Stichting Historisch Museum NLR (Amsterdam), NLL Vergaderingen Bestuur, Vol. 1940–1945.

105 Verkort Verslag over het jaar 1941 van de Stichting 'Nationaal Luchtvaart-Laboratorium' (NLL), Amsterdam 1942, 6, Stichting Historisch Museum NLR (Amsterdam).

106 Verkort Verslag over het jaar 1941 van de Stichting 'Nationaal Luchtvaart-Laboratorium' (NLL), Amsterdam 1942, 6–7, Stichting Historisch Museum NLR (Amsterdam).

107 Stichting NLL: Voorloopige resultatenrekening over het jaar 1942, Stichting Historisch Museum NLR (Amsterdam), NLL Vergaderingen Bestuur, Vol. 1940–1945.

108 JHdH/RV (Stichting NL): Resultatenrekening over het jaar 1943, Stichting Historisch Museum NLR (Amsterdam), NLL Vergaderingen Bestuur, Vol. 1940–1945.

109 Verslag over de jaaren 1944 en 1945 van de Stichting Nationaal Luchtvaart-Laboratorium (NLL), Amsterdam 1946, 21, Stichting Historisch Museum NLR (Amsterdam).
110 Verslag over de jaaren 1944 en 1945 van de Stichting Nationaal Luchtvaart-Laboratorium (NLL), Amsterdam 1946, 22–3, Stichting Historisch Museum NLR (Amsterdam).
111 SL/dM: Notulen van de elfde vergadering van het bestuur van de Stichting 'Nationaal Luchtvaartlaboratorium' gehouden op 26 March 1943 te Amsterdam, 7, Stichting Historisch Museum NLR (Amsterdam), NLL Vergaderingen Bestuur, Vol. 1940–1945.
112 This claim is based on the personal account written in 1983 by A. van der Neut, 'Het Nationaal Luchtvaartlaboratorium in oorlogstijd', Stichting Historisch Museum NLR (Amsterdam). The memoirs of van der Neut are published in: Stichting Historisch Museum NLR. Niewsbrief 23, maart 2007, 10–11; Niewsbrief 24, juni 2007, 10–11; and Niewsbrief 25, augustus 2007, 8–10.
113 Van der Neut 1983, 11.
114 Van der Bliek 1994, 30.
115 Budraß 1998, 244f.
116 O. Groehler, *Geschichte des Luftkriegs 1910 bis 1980* (Berlin: Militärverlag der DDR, 1990), 192–202.
117 J. H. Greidanus and A. I. van de Vooren, 'Berechnung der Flatterschwingungen beschrieben in den Versuchsberichten Nr. 934/13 und 933/29 der AVA. Auftrag V. 232.061, NLL report V.1284', 17 October 1942, Archiv Deutsches Museum (Munich), ZLDI, NLL V.1284.
118 Attached note by Hans Georg Küssner (AVA – Institut für instationäre Strömungen), 13 January 1943, Ibid.
119 J. H. Greidanus and A. I. van de Vooren, 'Berechnung der Flatterschwingungen beschrieben in den Versuchsberichten Nr. 934/13 und 933/29 der AVA. Auftrag V. 232.061, NLL report V.1284', 17 October 1942, Archiv Deutsches Museum (Munich), ZLDI, NLL V.1284.
120 Ibid., 3.
121 Ibid.
122 Betz to Lorenz (Reichsluftfahrtministerium - Technisches Amt LC 1), Betr.: Versuchsarbeiten in Paris-St.Cyr und in Amsterdam, 1 March 1941, DLR Archives, GOAR 1000, fot. 143–4.
123 Lorenz to Betz, Betr.: Versuchsarbeiten in Paris-St.Cyr und in Amsterdam - AVA Schreiben B.183/41 vom 1 March 1941, 19 March 1941, DLR Archives, GOAR 1000, fot. 142.
124 H. J. Ebert, J. B. Kaiser and K. Peters, *Willy Messerschmitt - Pionier der Luftfahrt und des Leichtbaues. Eine Biographie* (Bonn: Bernard & Graefe, 1992), 115.
125 Ebert, Kaiser and Peters 1992, 134.
126 Ibid., 119.
127 Küssner to Jennissen (Reichsluftfahrtministerium), Betr.: Forschungarbeiten in Amsterdam, 25 June 1941, DLR Archives, GOAR 2729, fot. 502.
128 Ibid., fot. 502–3.
129 AVA-Auftrag LA 10013 - van der Neut: Versuche an einem Außenflügel einfacher Konstruktion zur Bestimmung der Versteiffestigkeit in ihrer Abhängigkeit vom Verdremoment. NLL Bericht S.259, 31.8.1942, 2, DLR Archives, GOAR 363, fot. 4.
130 Käufl to AVA (telex), 28.6.1941, DLR Archives, GOAR 2729, fot. 501.
131 AVA-Auftrag LA 10013 - van der Neut: Versuche an einem Außenflügel einfacher Konstruktion zur Bestimmung der Versteiffestigkeit in ihrer Abhängigkeit vom Verdrehmoment. NLL Bericht S.259, 31.8.1942, DLR Archives, GOAR 363, fot. 2–44.
132 Direktor NLL to Focke Wulf GmbH Bad Eilsen, Betr.: NLL-Berichte S.259, 267, 272 und 275, 31.5.1944, DLR Archives, GOAR 2729, fot. 409.

133 Sicherheitspolitzei Aussenstelle Amsterdam to SS-Standartenführer Harster (Den Haag), Betr.: Fokker-Werke, 12 June 1941. Appendix: Vernehmungsniederschrift Friedrich Seekatz, geb. 15.12.1886 (Direktor der Fokker-Werke), Nederlands Instituut voor Oorlogsdocumentatie (NIOD), 77/1365.
134 Befehlshaber der Sicherheitspolizei und des SD für die besetzten niederländischen Gebiete: Notice about a phone call from SS-Gruppenführer Rauter, 6 May 1941, NIOD 77/1365.
135 R. B. Birn, 'Hanns Rauter. Höherer SS- und Polizeiführer in den Niederlanden', R. Smelser and E. Syring (eds), *Die SS. Elite unter dem Totenkopf. 30 Lebensläufe* (Paderborn: Schöningh, 2000), 408–17; and N. K. C. A. in 't Velt (ed.) *De SS en Nederland. Documenten uit SS-Archieven. 1943–1945* (Nijhoff:'s-Gravenhage 1976), 98–145.
136 SS-Sturmbannführer Lages (Befehlshaber der Sicherheitspolizei und des SD für die besetzten niederländischen Gebiete - Aussenstelle Amsterdam) to SS-Standartenführer Dr. Harster, 8 May 1941, NIOD 77/1365.
137 Lages to SS-Sturmbannführer Dr Hammer (Befehlshaber der Sicherheitspolizei und des SD für die besetzten niederländischen Gebiete - Den Haag), 9 May 1941, NIOD 77/1365.
138 Der Befehlshaber der Sicherheitspolizei und des SD für die besetzten niederländischen Gebiete, Vermerk. Betr. Fokker-Werke, 10 May 1941, NIOD 77/1365.
139 Käufl (Paris) to Betz, 25 July 1941, DLR Archives, GOAR 2729, fot. 494–5.
140 Pleines (Fokker Amsterdam) to Lorenz (Forschungsführung Reichsluftfahrtministerium), 4 February 1943, DLR Archives, GOAR 2729, fot. 435.
141 Ibid., fot. 436.
142 Lorenz (Forschungsführung Reichsluftfahrtministerium) to Engelbrecht (AVA), 18 Febraury 1943, DLR Archives, GOAR 2729, fot. 434.
143 Aktenvermerk über die Besprechungen der Übergabe des Windkanals Hispano-Suiza an die DVL, gez. Käufl, Seewald, Gezlaff, 11 February 1943, DLR Archives, GOAR 1033, fot. 62.
144 Käufl to Seewald (Forschungsführung), 2 March 1943, DLR Archives, GOAR 2729, fot. 441.
145 Ibid., fot. 442–3.
146 Ibid.
147 Ibid., fot. 443.
148 Käufl, Aktenvermerk über die Besprechungen der Eingliederung des NLL, A'dam am 26. –27.3.1943 in Göttingen, 31 March 1943, DLR Archives, GOAR 2729, fot. 55–6.
149 Verslag over de jaaren 1944 en 1945 van de Stichting Nationaal Luchtvaart-Laboratorium (NLL), Amsterdam 1946, 10, Stichting Historisch Museum NLR (Amsterdam). This conviction is also cited by Van der Bliek 1994, 31.
150 Betz to Pleines (Amsterdam), 29 March 1943, DLR Archives, GOAR 2729, fot. 438.
151 Verkort Verslag over het jaar 1943 van de Stichting 'Nationaal Luchtvaart-Laboratorium' (NLL), Amsterdam 1944, 7, Stichting Historisch Museum NLR (Amsterdam).
152 Verslag over de jaaren 1944 en 1945 van de Stichting Nationaal Luchtvaart-Laboratorium (NLL), Amsterdam 1946, 9, Stichting Historisch Museum NLR (Amsterdam).
153 Käufl to Betz, 28.4.1944, DLR Archives, GOAR 1033, fot. 11.
154 L. de Jong, *Het Koninkrijk der Nederlanden in de Tweede Wereldoorlog. Mei 1943 - Juni 1944*, Vol. 7.1 – 7.2 ('s-Gravenhage: Nijhoff, 1976), 873; and L. de Jong, *Het Koninkrijk der Nederlanden in de Tweede Wereldoorlog. Het laatste jaar* ('s-Gravenhage: Nijhoff, 1980), 96. In both cases de Jong misdates the arrest of Chaillet to the autumn of 1943. For documents and testimonies of former members of the Nationaal Comité see: *Enquêtecommissie Regieringsbeleid 1940–1945. Verslag houdende de uitkomsten van het onderzoek. Deel 4 C-II - De Nederlandse Geheime Diensten te*

Londen de verbindingen met het bzette gebied. Verhoren ('s-Gravenhage: Staatsdrukkerij en Uitgeverijbedrief, 1950), 1016–17, 1200–1; *Enquêtecommissie Regieringsbeleid 1940–1945. Verslag en Bijlagen. Deel 7 A en B - Leiding en voorlichting aan amtenaren en burgers in de bezette gebieden. Het contact met en de politiek ten aanzien van de verzetbeweging in Nederland. Verslag en Bijlagen,* ('s-Gravenhage: Staatsdrukkerij en Uitgeverijbedrief, 1955), 231–8; and *Enquêtecommissie Regieringsbeleid 1940–1945. Verslag en Bijlagen. Deel 7 C - Leiding en voorlichting aan amtenaren en burgers in de bezette gebieden. Het contact met en de politiek ten aanzien van de verzetbeweging in Nederland. Verhoren* ('s-Gravenhage: Staatsdrukkerij en Uitgeverijbedrief, 1955), 82, 568–9, 685, 710–11, 716.

155 Käufl to Betz, 28 April 1944, DLR Archives, GOAR 1033, fot. 11.

156 Letter from the archives of the Gedenkstätte und Museum Sachsenhausen to the author, 5 December 2006. Information according to the Häftlingsnummernliste, Archiv der Gedenkstätte Sachsenhausen, D 1 A/1025, p. 096 and Häftlingsnummernliste, D 1 A/1222, p. 27 back side.

157 For the biography of Chaillet see the web site: http://www.dutchsubmarines.com/men/commanding_officers/1418_j_l_chaillet.htm (last accessed 7 April 2007).

158 Verslag over de jaaren 1944 en 1945 van de Stichting Nationaal Luchtvaart-Laboratorium (NLL), Amsterdam 1946, 15, Stichting Historisch Museum NLR (Amsterdam).

159 Ms Bé Wenting, born in 1913, had been working for the NLL as a secretary since 1933. See: Personal-Liste des NLL, undated (ca. 1940), DLR Archives, GOAR 2729, fot. 577.

160 Verslag over de jaaren 1944 en 1945 van de Stichting Nationaal Luchtvaart-Laboratorium (NLL), Amsterdam 1946, 9–10, Stichting Historisch Museum NLR (Amsterdam).

161 Verslag over de jaaren 1944 en 1945 van de Stichting Nationaal Luchtvaart-Laboratorium (NLL), Amsterdam 1946, 14; and Personal-Liste des NLL, undated (ca. 1940), DLR Archives, GOAR 2729, fot. 579. J. Frits Hengeveld was arrested on 7 April 1944, cf. Koning (NLL) to ?, 18 April 1944, Stichting Historisch Museum NLR (Amsterdam), pers. dossier J. Frits Hengeveld.

162 Personal-Liste des NLL, undated (ca. 1940), DLR Archives, GOAR 2729, fot. 576. Verslag over de jaaren 1944 en 1945 van de Stichting Nationaal Luchtvaart-Laboratorium (NLL), Amsterdam 1946, 16, Stichting Historisch Museum NLR (Amsterdam). See also: Van der Bliek 1994, 31.

163 Teus van Oosterom, Bericht, 13 July 1945, Stichting Historisch Museum NLR (Amsterdam), pers. dossier Abraham Alfred Spitz.

164 Ibid.; and Mon Hartog (Centraal Registratiebureau voor Joden) to Nationaal Luchtvaartlaboratorium, 26 October 1945, Stichting Historisch Museum NLR (Amsterdam), pers. dossier Hartog Groen.

165 Verklaring voor bezolding van uit hun functie (nog niet) ontheven personen van joodsche bloede, wg. Groen - Chaillet 29 January 1941, Stichting Historisch Museum NLR (Amsterdam), pers. dossier Hartog Groen.

166 Chaillet to Blackstone (Stichting NLL), 22 January 1941, Stichting Historisch Museum NLR (Amsterdam), pers. dossier Hartog Groen.

167 Verslag over de jaaren 1944 en 1945 van de Stichting Nationaal Luchtvaart-Laboratorium (NLL), Amsterdam 1946, 10, Stichting Historisch Museum NLR (Amsterdam).

168 For the epidemic hunger as a result of the German occupation policy see: Z. Stein, *Famine and Human Development. The Dutch Hunger Winter of 1944–1945* (New York: Oxford University Press, 1975); and H. A. Van der Zee, *The Hunger Winter. Occupied Holland, 1944–1945* (Lincoln, Neb.: University of Nebraska Press, 1998).

169 In total a turnover of 40 tons of foodstuff was reached. Verslag over de jaaren 1944 en 1945 van de Stichting Nationaal Luchtvaart-Laboratorium (NLL), Amsterdam 1946, 13, Stichting Historisch Museum NLR (Amsterdam).

170 Verslag over de jaaren 1944 en 1945 van de Stichting Nationaal Luchtvaart-Laboratorium (NLL), Amsterdam 1946, 10, Stichting Historisch Museum NLR (Amsterdam).
171 Ibid., 12.
172 Ibid., 9.
173 Ibid., 14.
174 Ibid., 11, 14.
175 Ibid., 9.

10 Masa Takeuchi and his involvement in the Japanese nuclear weapons research programme

Masakatsu Yamazaki

There were two nuclear weapons projects in Japan during World War II: an army programme called 'Ni-go Research' and a much smaller 'F-Research' of the navy.[1] 'Ni-go Research' was directed by Yoshio Nishina, a leading nuclear physicist at the Riken, the Institute of Physical and Chemical Research. The experimental physicist Masa Takeuchi (1911–2001), a member of Nishina's nuclear research group, became one of the key figures of the army project. This paper will take Takeuchi's case as an example of Japanese scientists involved in wartime research during World War II. His case illustrates how Nishina's nuclear group adapted for survival to an extraordinary research environment in wartime.

Takeuchi's Early Work on Cosmic Rays

Masa Takeuchi entered the Applied Chemistry Department of the Tokyo Higher Technical School just before it became the Tokyo Technical College in 1929.[2] The college is now called the Tokyo Institute of Technology. He graduated in 1931 and became a research member of the radioactivity measurements team at Nishina's laboratory. At the time Nishina was building a large nuclear research group with a broad research perspective, including theoretical research, basic nuclear experiments and the application of radioactive substances to biological and medical areas. Takeuchi was given charge of the construction of a Wilson cloud chamber of 40 cm diameter accompanied by a large electromagnet (Figure 10.1). The chamber stood vertically, which enabled Takeuchi to take pictures easily. As there was no powerful electric supply facility in the Riken at the time, Takeuchi and his colleague Torao Ichimiya used a direct current generator used to power submarine batteries as the source of electricity for the electromagnet, so the experiments were done at the Yokosuka Naval Arsenal near Yokohama. They started their experiments at the arsenal in the spring of 1936. One year later they successfully obtained the energy spectrum of cosmic rays. However, their result merely confirmed P. M. S. Blackett's observation, which was published earlier the same year,[3] and their experiments were not publicly reported.[4,5]

In April 1937, Niels Bohr visited the Riken. He suggested to Nishina putting a lead plate into the cloud chamber to estimate the masses of particles in cosmic rays. Nishina thought it worthwhile to execute this idea, particularly since various

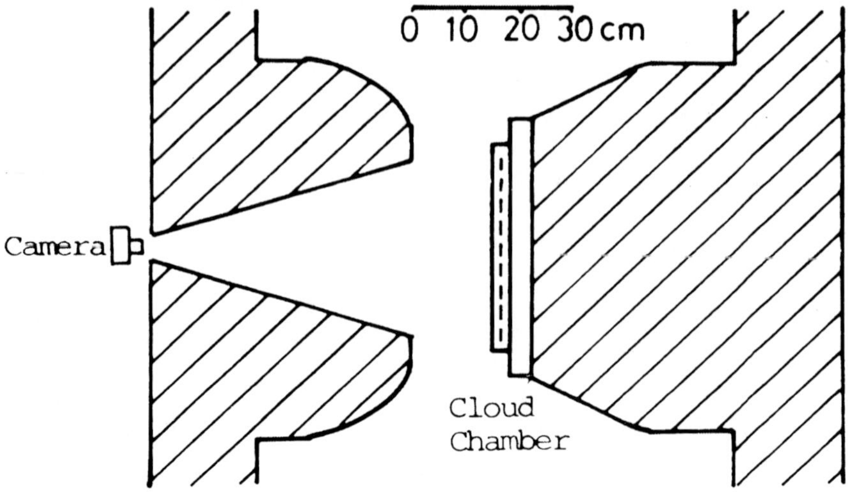

Figure 10.1 The cross-section of the electromagnet and the cloud chamber.[6]

experimental physicists had recently observed new particles of a mass heavier than that of an electron and lighter than that of a proton.[7] The observations suggested that these particles would be the mesons that the Japanese theoretical physicist Hideki Yukawa had predicted in 1931.[8] Theoretical physicist Minoru Kobayashi of the Nishina group helped Takeuchi with the calculation of the energy loss of charged particles of various mass by the ionisation effect in the lead plate. Takeuchi could then identify the mass of a particle by comparing the calculated energy loss and the particle's momentum change that he observed in the chamber. By the summer of 1937, Takeuchi successfully obtained clear pictures of the track of mesons from cosmic rays (these mesons were, in fact, later found not to be Yukawa's mesons – pions – but their daughter particles – muons) and estimated the mass at around 1/10 of that of a proton.[9,10] Nishina published an article on his group's discovery in a Japanese journal in September 1937 (Figure 10.2).[11] His joint paper in English with Takeuchi and Ichimiya appeared in the *Physical Review* as an independent report of the discovery of mesons in December 1937, seven months after the first announcement of the experimental discovery of mesons by Seth Neddermeyer and Carl D. Anderson.[12] These experimental achievements encouraged Yukawa and other Japanese theoretical particle physicists. Yukawa was to receive the Nobel Prize for physics in 1949.

Takeuchi's achievements went further. By the end of 1938, he observed a single track that suggested the existence of an unknown particle that had about one-half of the mass of a proton. He reported his finding at a colloquium at the Riken but, unfortunately, never published his results (Figure 10.3). He believed that it was possibly the first evidence of K-mesons (or kaons).[13] Further results of Nishina's group concerning the mass of mesons were published in January 1939.[14]

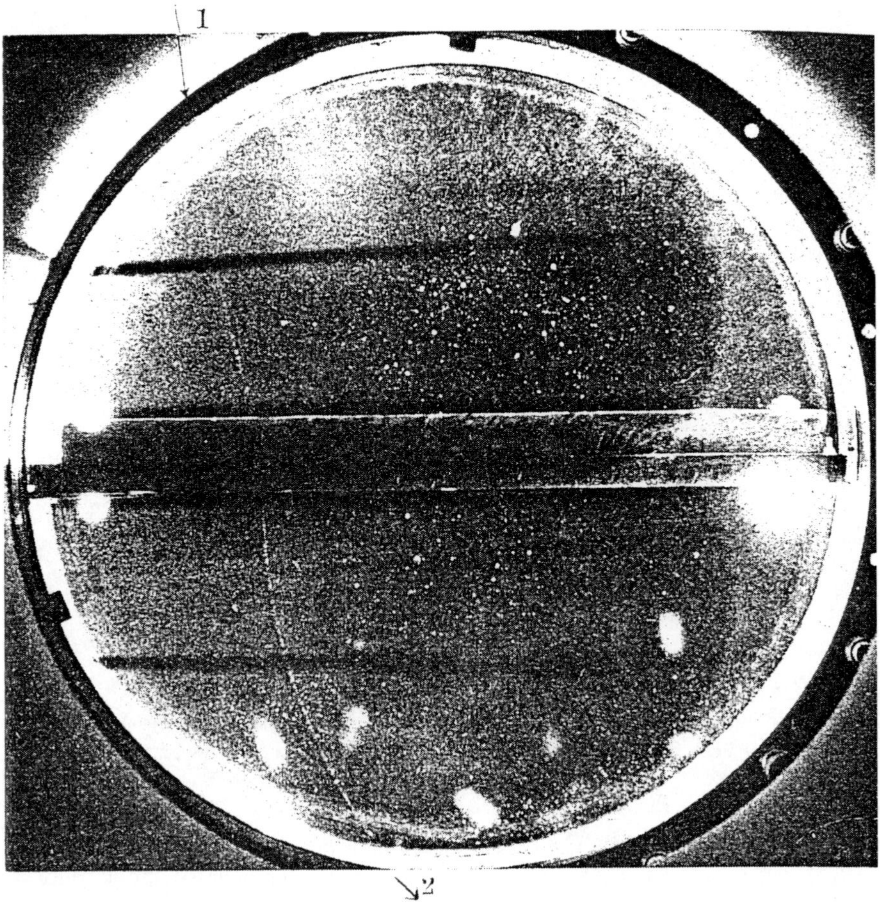

Figure 10.2 A picture of the meson's track from 1 to 2.[15]

Takeuchi was developing a much larger cloud chamber when his cosmic ray research was suddenly interrupted in December 1942, a year after the Pearl Harbor attack. Nishina asked Takeuchi to stop his cosmic ray experiments and instead to take on responsibility for uranium condensation.

Nishina's Nuclear Weapons Project and the Adaptation of Nishina's Group to the Wartime Research Environment

About six months before the Pearl Harbor attack Nishina was asked by the army to take charge of the nuclear weapons research. Masatoshi Ohkochi, director of the Riken, received the official commission from Takeo Yasuda, director of the Army Aeronautical Department's Technical Research Institute and later director

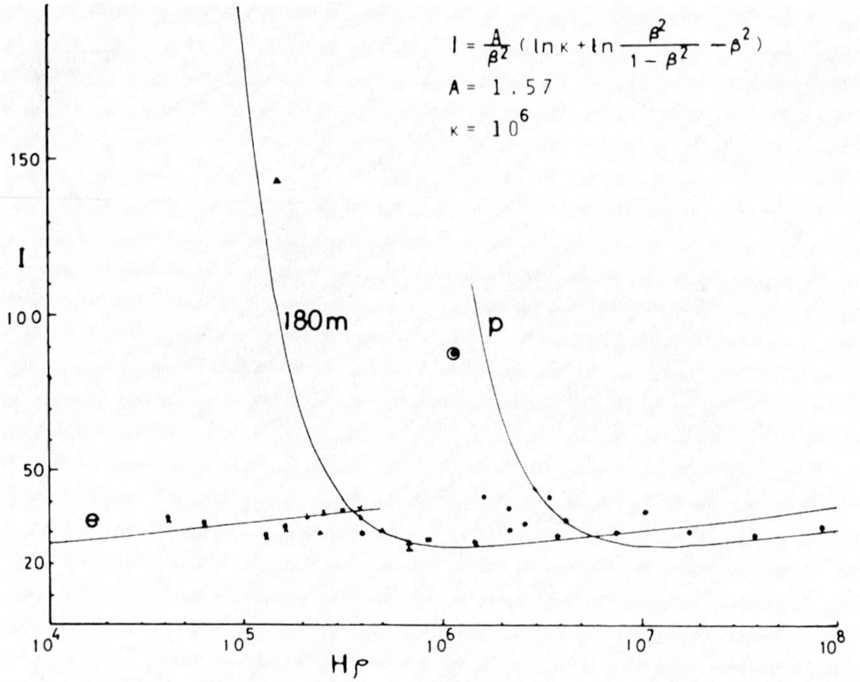

Figure 10.3 Takeuchi's observation of the momentum and ionisation density from particle tracks. The circle at the centre indicates the medium mass particle with one-half of the mass of a proton.[16]

of the Army Aeronautical Department, which was a chief patron of the army nuclear weapons project.[17] However, Nishina was less enthusiastic about nuclear weapons research even after the Pearl Harbor attack. In December 1941, he told his colleagues that they should continue their basic research.[18] Around that time, Nishina had asked Kunihiko Kigoshi, a chemist just graduated from Tokyo Imperial University, to join the project to make uranium hexafluoride. Nishina told him, 'If you work on uranium hexafluoride, you will not have to become a soldier'. Most interestingly, Nishina added, 'We shall go about this slowly'.[19]

Towards the end of 1942, however, Nishina's attitude was changing. As the war became more unfavourable for Japan, Nishina began to emphasise the importance of applied science more than that of basic science. In December 1942, Nishina told his colleagues, 'We must work for our country'.[20] Nishina then asked Masa Takeuchi and Hidehiko Tamaki to take responsibility for uranium-235 separation and the theoretical study of uranium nuclear chain reactions, respectively.[21]

During the war Takeuchi could not understand why Nishina 'ordered' him to undertake uranium condensation works.[22] It is possible that Nishina was afraid that Takeuchi would easily be drafted into military service at the front because cosmic ray experiments would be considered as having nothing to do with the nuclear weapons project. Nishina could easily persuade the military authorities that his other nuclear research programmes, such as cyclotron experiments, were parts of the weapons project. Takeuchi's cosmic rays research, however, would have required further explanations and this might be the reason that Nishina 'ordered' Takeuchi to undertake research directly related to nuclear weapons research.

Table 10.1 shows how Nishina's nuclear group was rearranged in the war period. Some of the scientists in his group could continue to do their own 'basic' research. Given their cases, one can say that the war provided a new opportunity for some scientists to advance their academic careers. But others had to abandon their previous studies to help with military research such as radar research.

A remarkable case is that of Shinichiro Tomonaga, a future Nobel Prize laureate for physics, who introduced the theoretical study of ultrashort-wave circuits for radar at the Navy Technical Research Institute and successfully developed his theory of ultra-short wave circuits. This situation fits into what the historian of science Peter Galison calls 'trading zone'. In his book *Image and Logic* Galison suggests that on some occasions collaboration between different scientific cultures or traditions results in fruitful achievements. He calls such a kind of collaboration 'trading zone'.[23] One can say that the trading zone between scientists and engineers worked well in the case of Tomonaga.[24] Tomonaga applied Heisenberg's S-matrix theory to his study of the complex wave circuits of various elements such as wave guides, cavity resonators or electromagnetic horns. In August 1944, he succeeded in formulating a general theory of them. This theory was what the engineers of the Navy Technical Research Institute had been looking for in order to design ultra-short wave circuits.

The involvement of these scientists in the war effort has been a controversial ethical issue in Japan, still generating debate. But if we put aside the ethical issue for a moment in order to avoid the emotional tension involved, we may be able to look at their behaviour like biologists, as if it were a natural phenomenon. Just as

Table 10.1 The wartime reorganisation of the Nishina research group

Nishina research group in 1936
Quantum theory (Nishina, Tomonaga, Tamaki and Kobayashi)
Cosmic rays (Nishina, Ishii, Takeuchi, Asano, Sekido and Ichimiya)
Artificial radioactivity (Nishina, Watanabe and Takeuchi)
Research on neutrons (Nishina, Tamaki, Yamazaki, Niima and Sugimoto)
Wartime arrangement of the group
Radar research (Tomonaga: theory of ultra-short wave circuits)
Cyclotron and radioactivity (Nishina, Yamazaki, Sugimoto and Tamaki)
'Ni-go research (nuclear weapons)' (Nishina, Tamaki [theoretical study], Kigoshi [chemical processing], Takeuchi [thermal diffusion], Yamazaki [assisted with physical tests])

animals and plants survive by adapting to severe environmental situations with their own strategies, Nishina's group was struggling to find a way to survive in an 'extraordinary' research environment of the war.

Takeuchi's Involvement in Wartime Nuclear Development

In March 1943, after several meetings, the Nishina group finally decided to adopt the thermal diffusion method to condense uranium-235. Based upon this decision and theoretical calculations by Tamaki, Nishina submitted his feasibility report to the army in June 1943. The report stated that:

1. the utilisation of nuclear energy by means of uranium fission is quite possible;
2. a minimum amount of 10 kg of 10 per cent enriched uranium is necessary for one explosion;
3. copper is stable for fluorine gas, but needs to be experimentally tested for uranium hexafluoride.[25]

The army soon approved Nishina's project as a high-priority project named 'Ni-go Research'. 'Ni' was taken from the first Japanese syllable of Nishina's family name. The approval of this project by the army enabled Takeuchi to obtain the materials he needed, copper pipes for example, although some of these materials were of strategic importance to the military.

Two months later, Takeuchi obtained all the parts and materials for the construction of test equipment five metres in height (Figure 10.4). By July 1944, Takeuchi had completed the equipment and started the condensation experiments with a small amount of the uranium hexafluoride that Kigoshi manufactured. The substance produced using this equipment was tested by Fumio Yamazaki with a small cyclotron. However, no positive result was produced by February 1945. The air raid on the night of 13 April destroyed the thermal diffusion equipment and consequently Takeuchi's research programme came to an end.[26] In June, Nishina reported to the army that his project at the Riken had to be terminated.[27] Nishina then sent Takeuchi to the Navy Technical Research Institute where he helped with radar research until the end of the war.[28]

Conclusion

After the war, Takeuchi never again engaged in thermal diffusion research but taught students of the faculty of education of the Yokohama National University as a physics professor. Even long after the war, Takeuchi was still wondering why Nishina engaged in nuclear weapons research during the war and why he 'ordered' Takeuchi to undertake the uranium condensation works. Unfortunately, Nishina died in 1951 while Japan was still under Allied occupation (1945–1952) without providing an answer. It is understandable that Nishina had to be careful when talking about his wartime experiences in the occupation period as this was a sensitive issue at the time. Takeuchi did not dare to ask Nishina such questions.

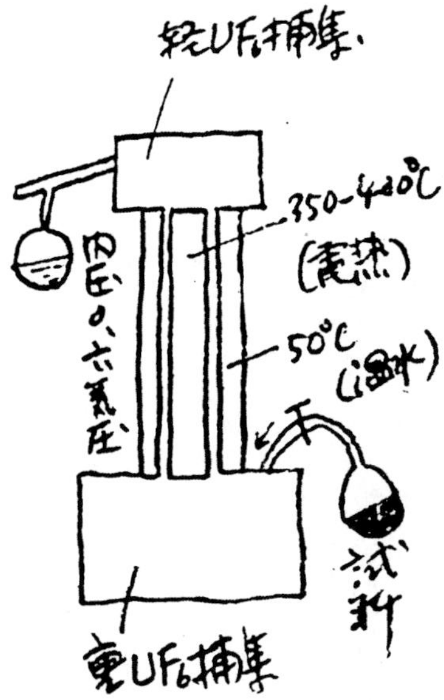

Figure 10.4 This picture of Takeuchi's thermal diffusion equipment appeared in Nishina's status report to the army in November 1944.[29]

Takeuchi left two essays on his involvement in the wartime nuclear weapons project, in which he tried to make clear why Nishina accepted a wartime research programme from the army. According to Takeuchi, Nishina's true research intention was the construction of a large cyclotron. However, he could not pursue this alone without substantial funding and he therefore reluctantly decided to undertake the nuclear weapons research, for which the construction of a large cyclotron was necessary. Otherwise the army would have withdrawn its research grant for Nishina.[30]

On the other hand, Takeuchi never found a clear answer to why Nishina involved him in the nuclear programme. But as has been mentioned above, it is likely that this was part of the adaptation of Nishina's nuclear group to an 'extraordinary' wartime research environment.

Notes

1 For a recent comprehensive historical study, see K. Nagase-Reimer, W. Grunden, and M. Yamazaki, 'Nuclear Weapons Research in Japan during the Second World War', *Historia Scientiarum* 14-3 (2005), 201–40. It includes a historiography of Japanese wartime nuclear research as well as a bibliography of related articles and books.

2 M. Takeuchi, 'Uran bakudan kenkyu mukashi banashi [An Old Story of Uranium Bomb Research]', *Gijutu-shi* [*History of Technology*] 3 (2002), 1–53.

3 P. M. S. Blackett, 'Further Measurements of Cosmic Ray', *Proc. Roy. Soc.* 159 (1937), 1–31.

4 M. Takeuchi, 'Cosmic Ray Study in Nishina Laboratory', Y. Sekido and H. Elliot (eds) *Early History of Cosmic Ray Studies* (Dordrecht: D. Reidel Publishing Company, 1985), 137–45.
5 M. Takeuchi, 'Kiribako ni yoru uchu sen no kenkyu [Cosmic Ray Study with Cloud Chambers]', H. Ezawa and H. Tamaki (eds), *Nishina Yoshio: Nihon no gennsi kagaku no akebono* [*Yoshio Nishina: The Down of Japanese Atomic Science*] (Tokyo: Misuzu Shobo, 1991), 104–11.
6 Takeuchi 1985, 139.
7 J. C. Street and E. C. Stevenson, 'Penetrating Corpuscullar Component of the Cosmic Radiation', *Phys. Rev.* 51 (1937), 1005; S. Neddermeyer and C. D. Anderson, 'Note on the Nature of Cosmic-Ray Particles', *Phys. Rev.* 51 (1937), 884–6.
8 H. Yukawa, 'On the Interaction of Elementary Particles I', *Rroc. Phys.-Math. Soc. Jpn.* 17 (1931), 48–57.
9 The earliest announcement of the Nishina group's discovery of muon tracks can be seen in Nishina's letter to E. C. G. Stuckelberg dated 3 August 1937, R. Nakane, Y. Nishina, K. Nishina, Y. Yazaki and H. Ezawa (eds), *Nishina Yoshio ouhuku shokan shu* [*Correspondences of Yoshio Nishina*] (Tokyo: Misuzu Shobo, 2006), II, 596–601.
10 Takeuchi 1991.
11 Y. Nishina, 'Shin ryushi no hakken [Discovery of New Particles]', *Kagaku*, September (1937), 408–11.
12 Y. Nishina, M. Takeuchi and T. Ichimiya, 'On the Nature of Cosmic-Ray Particles', *Phys. Rev.* 52 (1937), 1198–9.
13 Takeuchi 1991, 109.
14 Y. Nishina, M. Takeuchi and T. Ichimiya, 'On the Mass of the Mesotron', *Phys. Rev.* 55 (1939), 585–6.
15 Nishina 1937.
16 Takeuchi 1985, 142.
17 T. Yasuda, 'Nihon ni okeru genshi-bakudan seizo ni kansuru kenkyu no kaiko [Review of Research for Production of Atomic Bomb in Japan]', *Genshiryoku Kogyo*, 1, no. 4 (1955), 44–7.
18 Yomiuri Shimbun-sha (ed.), 'Nihon no genbaku [Japanese Atomic Bomb]', *Showa-shi no tenno* [*The Emperor in Showa Period*], vol. 4 (Tokyo: Yomiuri Shimbun-sha, 1968), 164 (Mitsuo Taketani's account).
19 Yomiuri Shimbun-sha 1968, 102–4.
20 Yomiuri Shimbun-sha 1968, 86.
21 Nihon Kagakushi Gakkai [the History of Science Society of Japan] (ed.), 'Butsurigaku to senji kenkyu', *Nihon kagaku gijutsushi taikei* [*An Outline of the History of Japanese Science and Technology*], vol. 13 (Tokyo: Daiichi Hoki Shuppan, 1970), 446.
22 Nihon Kagakushi Gakkai 1970, 445.
23 P. Galison, *Image and Logic* (Chicago and London: The University of Chicago Press, 1997). In Chapter 9 Galison emphasises the importance of 'trading' between scientists and engineers in wartime projects such as radar research at the Radiation Laboratory of MIT.
24 The result of Tomonaga's research work on ultra-short wave circuits was published after the war. See S. Tomonaga, 'General Theory of Ultra-short Wave Circuits I, and II', *Journal of the Physical Society of Japan* 2, no. 6 (1947), 158–71, and 3, no. 1 (1948), 93–105.
25 Yasuda 1955, 45.
26 Yomiuri Shimbun-sha 1968, 102–4.
27 Y. Yamamoto, 'Nihon genbaku no shinso [True Story of Japanese Atomic Bomb]', *Daihorin*, 20/8 (1953), 6–40.
28 Yomiuri Shimbun-sha 1968, 208.
29 *Tonizo* Report, the archives office at the Riken.
30 Takeuchi 2002, 42.

11 The cyclotron and the war

Construction of the 60-inch cyclotron in Japan

Keiko Nagase-Reimer

Experimental nuclear research started at the end of 1920s. In 1932, in Berkeley, Ernest O. Lawrence invented a useful experimental instrument for this purpose, the cyclotron. A cyclotron is a particle accelerator that gives charged particles high energy. With these particles, physicists could observe various nuclear phenomena in the laboratory.

Scientists outside the United States – in Japan for example – tried to emulate Lawrence's invention and were keen to build such a scientific instrument. In 1937, the eminent experimental physicist Yoshio Nishina[1] (1890–1951) and his fellow researchers built a 26-inch cyclotron and in 1944 a 60-inch cyclotron at the Institute of Physical and Chemical Research (Riken). The construction of these cyclotrons, especially the large one, required a huge amount of money. The cost of the magnet alone was as much as 110,000 yen (roughly equivalent to 1.2 million euros today) and that was beyond the annual budget of the institute. For such a 'Big Science' project the financial support of the government and industry was essential. This paper will show how Nishina was able to carry out the large cyclotron project in a time of war, when financial support, material and manpower were generally in short supply.

Promoting Science for the Rise of Japan

During and after the First World War people in Japan became aware of the importance of scientific research in order to develop new weapons as well as to be economically successful. It was for these reasons that the promotion of scientific research started. For example, the Institute of Physical and Chemical Research, the 'Riken', was established in 1917 in order to promote science. It was financed by the government, business circles and the imperial family.[2]

Nishina re-joined this institute in 1929 after returning from Europe, where he had spent a year in Cambridge, a year in Göttingen and more than five years at Niels Bohr's institute in Copenhagen. Fascinated by the open research atmosphere he had experienced, Nishina introduced the 'Copenhagen spirit' to the Riken. In 1931, Nishina became the leader of a research group. In the same year Japan launched its attack on China. The Japanese standing army stationed in Manchuria, the north-eastern part of China, invaded the city of Fengtian (today's

Shenyang). A year later Japan established the nation of Manchuria and installed a puppet government.

In January 1935, Nishina opened the new Nuclear Research Laboratory within the Riken together with his colleague Nishikawa. This new laboratory was realised by the financial support of the Mitsui Hōonkai Foundation, the Tokyo Electric Light Company and the Japan Wireless Telegraph Company.[3] Such financial contributions from industry had at first glance nothing to do with the military situation. But in fact they were deeply related to the military actions against China, since they were only possible because of the economic boom based on the colonial exploitation and the rapidly expanding war industry.[4]

There were two main ideas at this new Nuclear Research Laboratory: one was the construction of a Cockcroft-Walton accelerator, of which Nishina's colleague Nishikawa was in charge; the other was the building of a cyclotron, for which Nishina was responsible. At first Nishina planned to build a 26-inch cyclotron, the 'small cyclotron'. Nishina obtained intellectual support from Lawrence in Berkeley for its construction, sending two of his assistants, Ryōkichi Sagane and Tameichi Yasaki, to Berkeley in the summer of 1935.[5] Sagane stayed for a year and gathered the knowledge and experience necessary for the construction of cyclotrons. In the spring of 1936, Nishina and his fellow researchers in Japan began the construction of the small cyclotron. A year later the construction was complete and the small cyclotron produced its first beam.[6] Apart from the United States, Japan became the first country in the world to operate a cyclotron successfully.[7]

Nishina and his fellow researchers obtained important results using this small cyclotron. Some of these results were published in the well-known journal *Physical Review*. This indicates that Nishina and his team had reached the research level of their Western colleagues. However, he was not satisfied with the small cyclotron. He started another project for the construction of an even larger cyclotron, with the aim of being able to produce a stronger beam.[8]

The Large Cyclotron Project

In May 1936, just as the construction of the small cyclotron was beginning, Nishina had already started working out how much money he would need for a 60-inch electromagnet for the large cyclotron.[9] Two months later Nishina asked Lawrence about the price of a 60-inch electromagnet in the United States.[10] Through these actions Nishina knew that producing such an electromagnet in Japan would cost more than ordering it in the United States.[11] In 1937, Nishina ordered an electromagnet from the United States through the Japanese company Mitsui. Regarding this purchase, Lawrence greatly assisted his Japanese colleague by finding a suitable maker, by taking part in price negotiation and by supervising the construction.[12] This electromagnet arrived in Japan in the spring of 1938, not as a whole, but in parts. On its arrival, these parts were assembled by one of the biggest Japanese shipbuilding companies Ishikawajima. The completed electromagnet, with a weight of 200 tons, was transported to the Riken in the early summer of 1938.[13]

Figure 11.1 Yoshio Nishina in front of the large cyclotron. (Picture courtesy of the Nishina Memorial Foundation.)

This large cyclotron project required a huge amount of money. The electromagnet alone cost 110,000 yen. Nishina was able to secure financial support from the Japan Society for the Promotion of Scientific Research (*Gaku-shin*).[14] This Society had been established in 1932 with an extremely large budget to contribute to the rise of Japan through the promotion of science. The strategy of the Society was first to promote cooperation between universities and research institutes and then to make a link between research activities on the one hand and industry and military on the other hand.[15] Here again, the connection with the military can be found.

In addition to the costs for the electromagnet, 180,000 yen were estimated for the construction of the large cyclotron.[16] It is not clear where Nishina obtained the financial support for this.[17]

When Nishina started the construction of the large cyclotron he was very optimistic and estimated that it would soon be working.[18] The instrument was assembled in 1939, but it did not accelerate particles. In June 1940, Nishina decided to

Figure 11.2 The large cyclotron under construction. (Picture courtesy of the Nishina Memorial Foundation.)

ask Lawrence for advice, since he had already been successful in getting his 60-inch cyclotron running in the summer of 1939.[19] Nishina sent three of his assistants, Tameichi Yasaki, Sukeo Watanabe and Takeo Iimori, to Berkeley in the summer of 1940. But it was no longer possible for these three young Japanese physicists to see the 60-inch cyclotron: the president of the university did not allow the Japanese to visit the laboratory since military research was conducted there.[20] During their stay in the United States, more exactly on 27 September 1940, the Rome-Berlin-Tokyo Axis was formed. The anti-Japanese tone became stronger in the United States and some of the American colleagues refused to talk to the Japanese visitors.[21] Nevertheless, thanks to Lawrence, the three Japanese physicists received a rough blueprint of his cyclotron as well as published papers describing the mechanism of acceleration. As soon as they had returned to Japan in November 1940 the remodelling of the large cyclotron was started.[22]

For various reasons, this remodelling proceeded very slowly. Firstly, there was no suitable high-frequency system available in Japan. In the circumstances of war, Japanese engineers were busy producing devices for weaponry and there

Figure 11.3 The large cyclotron under construction. (Picture courtesy of the Nishina Memorial Foundation.)

was no place to produce anything for experimental instruments such as cyclotrons.[23] Secondly, Nishina's team was no longer able to obtain support from its colleagues in the United States, on which it was highly dependent. When Yasaki asked Cooksey to send him more blueprints of detailed designs of the large cyclotron at the end of 1940, Cooksey was not in a position to meet the wishes of his Japanese colleague.[24] And when Japan started its war against the United States in December 1941 communication with Lawrence broke off completely. Thirdly, the material situation became worse and worse as Japan began to lose battles from the summer of 1942.

Despite the slowness of the remodelling, the large cyclotron finally succeeded in producing a strong beam in January 1944.[25] From the summer of 1944, it was available for experiments, but hardly any were done.[26] Because of the daily air raids that Tokyo had suffered from the end of 1944, researchers were busy evacuating. One such air raid in April 1945 destroyed the biggest part of the Riken's buildings. The small cyclotron was damaged, but the large one survived until the end of the war without any damage.[27]

Nuclear Weapons Research

In the early summer of 1940, Nishina had by chance met Lieutenant-General Takeo Yasuda, at the time director of the Technical Research Institute of the Army Aeronautical Department, on a train. In the course of this encounter Nishina told Yasuda about the possibility of building nuclear weapons and that he would undertake experimental research for such nuclear weapons. This was the beginning of one of the two nuclear weapons projects during World War II in Japan.[28] The official commission to Masatoshi Ōkōchi, director of the Riken, was issued in June 1941 and Nishina was appointed for this nuclear research. This secret Army project was called the 'Ni-gō' project.

Why did Nishina draw the attention of the Army to nuclear weapons? To find the answer, this author examined what was actually done in the Ni-gō project. Within the framework of the Ni-gō project the researchers of Nishina's group were engaged in five research themes: theory, separation of uranium-235, production of uranium hexafluoride, measurement of physical constants and analysis of separated isotopes.[29] Firstly, the theory team calculated the critical mass and examined the principle of a nuclear weapon. This principle, however, was not technically feasible.[30] Secondly, the method of thermal diffusion was chosen for the separation of uranium-235. Masa Takeuchi was in charge of thermal diffusion, though he was not a specialist in this field.[31] Thirdly, Kunihiko Kigoshi pursued the production of uranium hexafluoride. He worked alone until technical officers from the Army were sent to assist him in March 1944.[32] Finally, there were plans to measure physical constants and to analyse separated isotopes using cyclotrons. The remodelling of the large cyclotron belonged to these two themes. The word 'cyclotron' does not appear on the list of researchers of the Ni-gō project. But in the plan submitted to the Technology Board (*Gijutsu-in*) in February 1944 the research using the large cyclotron was mentioned as one of two main themes of the Ni-gō project.[33] The remodelling of the cyclotron had priority over other themes and was pursued energetically.[34] These facts together points to the suggestion that Nishina drew the attention of the Army to nuclear weapons research in order to carry out the large cyclotron project. His 'technical sweet' was obviously the construction of the large cyclotron, not the development of nuclear weapons.

Nishina did not only push ahead with his large cyclotron project because it was his 'technical sweet'. He was convinced that the large cyclotron would contribute to the progress of science in Japan. He felt responsible for the development of science in Japan, even during the war. His strong sense of responsibility, mixed with the obligation to his mother country during the war, pushed him to promote the large cyclotron project.[35]

There were some advantages to working on the large cyclotron within the framework of the secret military project Ni-gō. First of all, there was the financial advantage. In total, 2,700,000 yen were offered for the Ni-gō project and Nishina could use this money in part for the large cyclotron.[36] In addition, in 1942 the Technology Board (*Gijutsu-in*) offered 90,000 yen for the construction of the large cyclotron and the research using it because the Board saw this project, being part of the nuclear weapons research project, as militarily high potential.[37] Secondly, Nishina could acquire the material he needed, although in the final phase of the war he could not get enough because of the general shortage of materials. Thirdly, Nishina could keep some of his fellow researchers away from other projects or from being drafted into the military.

Table 11.1 The list of researchers of the Ni-gō project

Group	*Team*	*Theme*	*Researcher*	*Place*
Plan (Colonel Arimori)	Plan	General plan	Major Koyama	Army Air Force Headquarters
Research (Yoshio Nishina)	Theory	Theoretical calculations	Hidehiko Tamaki Nobuyuki Fukuda	Riken
	Isotope separation 1	Isotope separation with thermal diffusion method	Masa Takeuchi First Lieutenant Saji Second Lieutenant Kittaka Second Lieutenant Hayashi Tamio Kobayashi	Riken
	Isotope separation 2		Major Suzuki First Lieutenant Kimoto Seinosuke Ozaki	Osaka Imperial University
	Chemistry	Chemical reactions and purification	First Lieutenant Ishiwatari Kunihiko Kigoshi First Lieutenant Sekihara First Lieutenant Wanibuchi Second Lieutenant Kawamura Second Lieutenant Funatsu	Riken
	Physics	Measurement of physical constants	Yoshio Nishina Keizō Niima Asao Sugimoto Eizō Tajima	Riken
	Analyse 1	Analysis of separated isotope	Yoshio Nishina Fumio Yamasaki First Lieutenant Kinoshita First Lieutenant Nakane Takahiro Namamoto	Riken
	Analyse 2		Tsunesaburô Asada Tsuyoshi Okuda Iichi Ogata	Osaka Imperial University
Materials (Satoyasu Iimori)	Materials	Search for materials and production of U_3O_8	Satoyasu Iimorio Shin Hata First Lieutenant Nakane Otokichi Nagashima	Riken

One interesting question is whether or not Nishina had a guilty conscience about his involvement in the project for weapons of mass destruction. This author does not know of any statement mentioning his own ethical attitude towards this. He was not confronted to reflect on his attitude by others, nor did he do so himself. During the war it was clear to him that Japan would never be able to develop nuclear weapons because; in the first place, there was not enough uranium.[38] So he never had to confront the issue of actually producing nuclear weapons or not.

Destruction of Cyclotrons

After the end of the war Nishina was full of hope that he could now concentrate on the large cyclotron. On 16 October 1945, Nishina sought the permission of the General Headquarters of the American occupational forces to use the two cyclotrons at the Riken for biological and medical research and the permission was granted the next day. However, the situation soon changed. On 10 November, the General Headquarters in Japan received the instruction from the Secretary of War in Washington to destroy all the cyclotrons at the Riken, at Kyoto Imperial University and at Osaka Imperial University. Following these instructions the four Japanese cyclotrons were destroyed on 24 November 1945. The Riken's cyclotrons were taken apart and thrown into the Gulf of Tokyo.[39]

In a written letter of protest against the destruction of cyclotrons Nishina wrote that the ones at the Riken had had nothing to do with the production of nuclear weapons.[40] What Nishina was claiming was more or less true: in fact, the large cyclotron was not used for nuclear weapons research. The small cyclotron was used for the analysis of separated isotopes, but mainly for biological and medical research. In order to obtain uranium-235 for a nuclear weapon Nishina would have had to convert the cyclotrons at the Riken into a mass spectrometer for uranium isotope separation, but such a suggestion was never made. Nevertheless, officially the large cyclotron project was part of the Ni-gō nuclear weapons project.

Hearing about the destruction of the Japanese cyclotrons, a great many scientists in the United States raised their voices in protest. A few weeks later Patterson, the Secretary of War, officially admitted that the instruction to destroy the Japanese cyclotrons had been a mistake.[41]

For Nishina, the destruction of the cyclotrons at the Riken meant the end of his research life. After the destruction he continued to be involved in science administration, became president of the Riken in 1946 and helped to reconstruct and democratise the Japanese scientific community. But he never returned to research himself. In January 1951, at the age of 60, Nishina passed away as a result of liver cancer.

Conclusion

By putting the large cyclotron project within the framework of the secret military nuclear weapons project 'Ni-gō', Nishina could continue the work on the large cyclotron until the end of the war. Under the name of the secret military project

Nishina could both secure the financial support and the material he needed and keep his fellow researchers away from other projects or from being drafted into the military. Despite such advantages the war had a negative impact on the project. Apart from shortages of materials in general, the communication between Nishina and Lawrence came to an end, communication on which Nishina was highly dependent. Other Japanese engineers were busy producing devices for weaponry, so there was no place to produce anything for experimental devices such as cyclotrons. The deciding and fatal influence of the war was above all the destruction of the cyclotrons.

In order to find financial sponsors Nishina tried to make his cyclotron project fit in with the political circumstances of the time. He made his large cyclotron project attractive by suggesting that this scientific instrument would serve basic research for the application of nuclear energy in the future. He did this without having a guilty conscience because he did not see any possibility at all of producing nuclear weapons in Japan during the war.

Notes

Abbreviations used:

EOL Ernest O. Lawrence Papers, The Bancroft Library, University of California, Berkeley.

NY1 Ryōhei Nakane, Yūichirō Nishina, Kōjirō Nishina, Yūji Yazaki and Hiroshi Ezawa eds., *Nishina Yoshio ōfuku shokanshū* [*Correspondences of Yoshio Nishina*] vol. I, Misuzu shobō, Tokyo (2006).

NY2 Ryōhei Nakane, Yūichirō Nishina, Kōjirō Nishina, Yūji Yazaki and Hiroshi Ezawa eds., *Nishina Yoshio ōfuku shokanshū* [*Correspondences of Yoshio Nishina*] vol. II, Misuzu shobō, Tokyo (2006).

NY3 Ryōhei Nakane, Yūichirō Nishina, Kōjirō Nishina, Yūji Yazaki and Hiroshi Ezawa eds., *Nishina Yoshio ōfuku shokanshū* [*Correspondences of Yoshio Nishina*] vol. III, Misuzu shobō, Tokyo (2007).

1 In this paper Japanese names are written according to the Western style (first given name and second surname).
2 T. Hiroshige, *Kagaku no shakaishi* [Social History of Science] (Tokyo: Chūō kōron sha, 1973), 92.
3 D.-W. Kim, 'Yoshio Nishina and two cyclotrons', *Historical Studies in the Physical and biological Sciences (HSPS)*, 36 (2), (2006), 248.
4 Kim 2006, 247; S. Hinokawa, 'Cyclotron Development at the Institute of Physical and Chemical Research in the 1930s', *TITech Studies in Science, Technology and Culture* 4 (2001), 16–17.
5 Nishina to Lawrence, 26 July 1935 (EOL, 030044:39); Yasaki (Berkeley) to Nishina, 11 October 1935 (NY1, 390–4).
6 Kim 2006, 247–50; Hinokawa 2001, 17–21.
7 C. Weiner, 'Cyclotrons and Internationalism: Japan, Denmark and the United States, 1935–1945', *Proceedings of the XIV International Congress of the History of Science, 1974* (2), Science Council of Japan (Tokyo, 1975), 353.

8 Kim 2006, 260; H. Nakayama, 'Toreisā to shokubutsu seiri no kenkyū [Research on tracer and physiology of plants]', H. Tamaki and H. Ezawa (eds), *Nishina Yoshio Nihon no genshi kagaku no akebono* (Tokyo: Misuzu shobō, 1991), 149.
9 Nishina to Hiroshi Toyota (Hitachi), 19 May 1936 (NY2, 444–5); Kim 2006, 250.
10 Nishina to Lawrence, 30 July 1936 (EOL, 029901:38).
11 Nishina to Sagane (Cambridge), 17 January 1937 (NK2, 524); E. Tajima, 'Riken no saikurotoron [Cyclotrons in the Riken]', Tamaki and Ezawa 1991, 121–2 .
12 Nishina to Lawrence 1 June 1937 (EOL, 029901:38).
13 Tajima 1991, 121–2; Kim 2006, 260—1; T. Tsuji, 'Riken jidai to nihon no butsurigaku [Riken Days and Physics in Japan]', Tamaki and Hiroshi 1991, 30.
14 Dai saikurotoron kensetsuhi [construction cost of the large cyclotron] April 1937 (NY2, 564–5); Nishina to Sagane (Cambridge University), 14 February 1937 (NY2, 537–41).
15 Hiroshige 1973, 121.
16 Dai saikurotoron kensetsuhi [construction cost of the large cyclotron] April 1937 (NY2, 564–5).
17 I thank Shizue Hinokawa for comments.
18 Nishina to G. Hevesy (Copenhagen), 28 August 1937 (NY2, 619–21); Hinokawa (ref. 2), 29.
19 Nishina to Lawrence, 26 February 1940 (EOL, 029901:38); Tsuji (ref. 12), 30.
20 Nishina to Lawrence, 15 June 1940 (EOL, 029901:38); Lawrence to Nishina, 22 Aug 1940 (EOL, 029901:38).
21 Yasaki (Columbia University, New York) to Nishina, 12 October 1940 (NY3, 949–52).
22 Yasaki (Berkeley) to Nishina, 1 August 1940 (NY3, 936–40); Toshio Takamine (Berkeley) to Nishina, 13 September 1940 (NY3, 941–2); Tajima 1991, 122–3; Nishina to Lawrence, 28 Nov 1940 (EOL, 029901:38).
23 Hinokawa 2001, 32.
24 Yasaki to Cooksey, 29 November 1940 (EOL, 030320:43); Cooksey to Yasaki, 2. January 1941 (EOL, 030320:43).
25 Tsuji 1991, 32; Tajima 1991, 124.
26 Kim 2006, 253, 263.
27 Tajima 1991, 124.
28 T. Yasuda, 'Nihon ni okeru genshi bakudan seizō ni kansuru kenkyūno kaiko [Retrospection of nuclear weapons research in Japan]', *Genshiryoku kōgyō*, 1(4) (1955), 44. The another project was the Navy's 'F-go'
29 See Table 11.1. Nihon Kagakushi Gakkai (ed.), *Nihon kagaku gijutsushi taikei* [compendium history of science and technology in Japan], 13 Butsuri Kagaku [Physics] (Tokyo: Daiichi hōki shuppan, 1970), 465.
30 K. Nagase-Reimer, W. E. Grunden and M. Yamazaki, 'Nuclear Weapons Research in Japan during the Second World War', *Historia Scientiarum*, 14 (3) (2005), 221–2.
31 See Yamazaki's paper in these Proceedings.
32 K. Nagase-Reimer, *Forschungen zur Nutzung der Kernenergie in Japan, 1938–1945*, (Marburg: Förderverein Marburger-Japan-Reihe, 2002), 65–6 .
33 Senji kenkyū jisshi keikaku, 2 February 1944 (NY3, 1078–9).
34 Nihon Kagakushi Gakkai 1970, 445.
35 S. Miyata, *Kagakusha tachi no jiyūna rakuen: Eikō no rikagaku kenkyūjo* [The Scientists' Paradise of Freedom: The Glorious Institute of Physical and Chemical research] (Tokyo: Bungei shunjū, 1983), 235–6; M. Taketani, 'Genshiryoku to kagakusha [Atomic energy and scientists]', *Genshiryoku to kagakusha* (Tokyo: Keisō shobō, 1968) (published in 1958 for the first time), 345.
36 Nagase-Reimer, Grunden and Yamazaki 2005, 221.
37 M. Yamazaki, 'The Mobilization of Science and Technology during the Second World War in Japan (in Japanese)', *Tōkyō kōgyō daigaku jinbun ronsō* 12 (1994), 175. Also

see H. Nagaoka to T. Hirai (Ministry of Commerce and Industry) 18 January 1941 (NY3, 975–7).

38 Nagase-Reimer 2002, 54; Y. Kawamura and M. Yamazaki, 'Butsuri kondankai to kyū nihon kaigun ni okeru kaku oyobi kyōryoku magunetoron kaihatsu [Physical Committee and the Japanese Navy's Project for Development of Atomic Energy and Powerful Magnetron during the Second World War]', *Kagakushi kenkyū*, 37 (1998), 165.

39 Tajima 1991, 125; 'Cyclotron smashing', *Life*, 19:26 (24 Dec. 1945), 26–7.

40 Nishina to D. MacArthur (Tokyo), 12 December 1945 (NY3, 1195–6).

41 D. MacArthur, *Reminiscences* (London: Heinemann, 1965), 286–7; Y. Nishina, 'A Japanese Scientist Describes the Destruction of His Cyclotrons', *Bulletin of the Atomic Scientists*, 3 (6) (1947), 145, 167; L. M. Groves, *Now It Can Be Told* (New York: Harper, 1963), 369–70; V. C. Jones, *Manhattan: The Army and the Atomic Bomb*, (Washington DC: Center of Military History, United States Army, 1985), 586–8. Also see Nagase-Reimer, Grunden and Yamazaki 2005, 232–3.

12 Forging a new discipline

Reflections on the wartime infrastructure for research and development in feedback control in the US, the UK, Germany and the USSR

Chris C. Bissell

Feedback control mechanisms – that is, systems whose behaviour is modified in some desired manner by manipulating them in response to the system output – have been known since ancient times.[1] Modelling them and predicting their behaviour, however, proved to be difficult, and became a major problem in the nineteenth century in connection with the control of steam and hydraulic turbines. While major strides were made in this field during the second half of the nineteenth and first half of the twentieth centuries,[2] it was during the Second World War that a discipline of feedback control began to emerge, using a range of design and analysis techniques to implement high-performance systems, in particular those for the control of anti-aircraft weapons.[3,4] In particular, World War II saw the coming together of engineers from a range of disciplines – electrical and electronic engineering, mechanical engineering, mathematics – and the subsequent realisation that a common framework could be applied to all the various elements of a complex control system in order to achieve the desired result. This approach, later known as 'systems engineering', was a major technological breakthrough.

A generic diagram of a simple feedback system with one input and one output is shown in Figure 12.1. The system output is compared to the loop input, and a controller is used to modify the behaviour of the process itself so as to cause it to change in a desired manner – or remain constant, depending on the application – under the influence of disturbances. An everyday example is the control of a heating system. The difference between the desired (input) and actual (output) temperature, measured by a thermostat, is used to control the boiler so as to maintain the desired temperature.

During the 1930s, high-performance, closed-loop, servomechanisms were designed for use with electromechanical simulators such as the differential analyser. In such servomechanisms the position of the device to be controlled was detected, and the value of any error between the actual and desired position was used to drive the mechanism in the desired direction. It was well known that under certain circumstances feedback loops became unstable, and a major task was to design the system such that the servomechanism responded rapidly, and with low steady-state error,

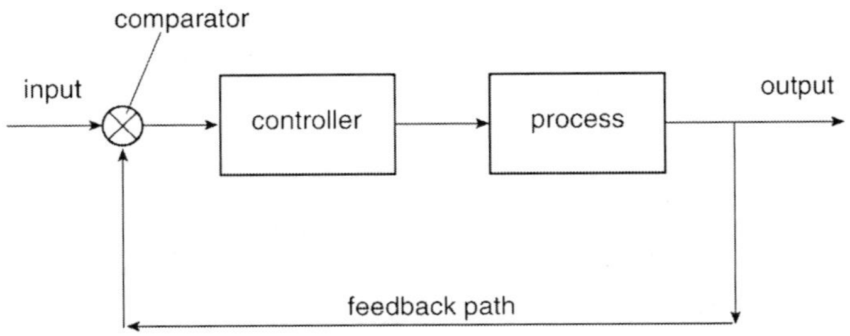

Figure 12.1 A feedback loop with one input and one output.

without becoming unstable or exhibiting undesirable oscillatory behaviour. Throughout the 1930s, individual developments in this field, while impressive, remained isolated. As David Mindell says of the US context: '... control engineering in 1940 remained local, tied to discrete engineering cultures. No conference had brought its practitioners together, no special publications were dedicated to feedback problems, no theory or textbooks solidified their common foundations'.[5]

The historical literature in English on the development of control engineering during the Second World War has tended to concentrate on the US and the UK. The specialist literature in German is often still surprisingly coy about the wartime period, and some of the activities during that time of major German contributors to feedback control are under-reported.[6] Until comparatively recently, much Russian historiography in science and technology has been tendentious and unreliable. Nevertheless, one surprising thing that emerges from a study of the state of feedback control at the end of World War II is that a good deal of the theory of the discipline had been developed independently in the various warring countries, even if the US (and, to a lesser extent, the UK) were well ahead in terms of actual design techniques and the implementation of functioning, high-performance, closed-loop systems. This chapter is an attempt to begin to put such wartime developments in an international context, concentrating on institutional and infrastructure aspects.

The so-called fire control problem was one of the major issues in military research and development at the end of the 1930s. While not a new problem, the increasing importance of aerial warfare meant that the control of anti-aircraft weapons took on a new significance. The state of the art at the beginning of the 1940s is illustrated in Figure 12.2. Figure 12.3 shows a more humorous version.

Under manual control, aircraft were detected, range was measured, prediction of the aircraft position at the arrival of the shell was computed, guns were aimed and fired. The system illustrated in Figure 12.2 could involve up to fourteen operators to handle the various manual tracking controls for radar dish orientation and gun directing. Clearly, automation of the process was highly desirable, and achieving this was to require detailed research into such matters as the dynamics of the servomechanisms driving the aiming of the gun, the design of controllers, and the statistics of tracking aircraft possibly taking evasive action.

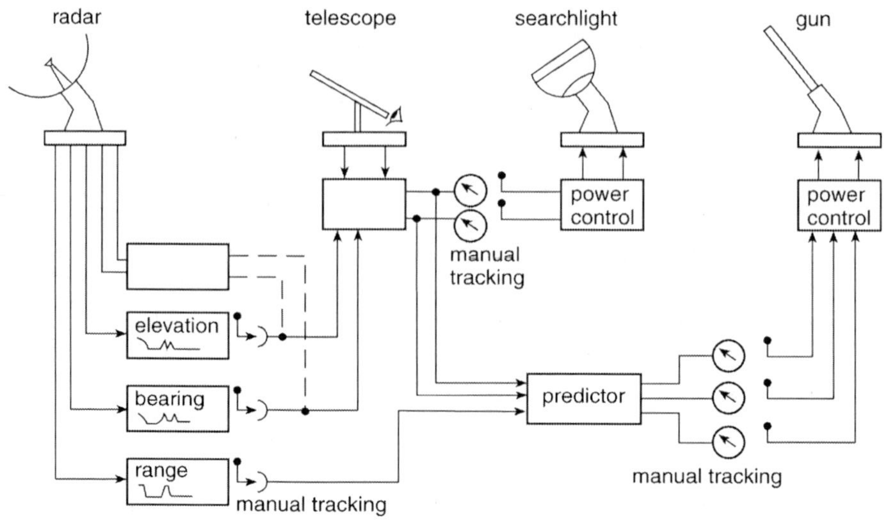

Figure 12.2 Manual fire control c. 1940 (redrawn from Bennett 1993).

The United States

Support for control systems development in the United States has been well documented.[7] The National Defense Research Committee (NDRC) was established in 1940 and incorporated into the Office of Scientific Research and Development (OSRD) the following year. Under the directorship of Vannevar Bush the new bodies tackled anti-aircraft measures, and thus the servo problem, as a major priority. Section D of the NDRC, devoted to Detection, Controls and Instruments, was the most important for the development of feedback control. Following the establishment of the OSRD the NDRC was reorganised into divisions, and Division 7, Fire Control, under the overall direction of Harold Hazen, covered the subdivisions: ground-based anti-aircraft fire control; airborne fire control systems; servomechanisms and data transmission; optical rangefinders; fire control analysis; and navy fire control with radar. This organisational structure brought about a significant change in the management of research contracts, over eighty of which were in the general fire control area.[8] However, there were still significant disputes and rivalries about which precise grouping was responsible for individual projects.

Government, industry and academia worked closely on the wartime agenda in the US, and three research laboratories were of prime importance, between them employing most of the American founding fathers of classical control. The Servomechanisms Laboratory at MIT brought together G. S. Brown, A. C. Hall, J. W. Forrester and others in projects that developed frequency-domain methods for control loop design for high-performance servos. Particularly close links were maintained with Sperry, a company with a strong track record in guidance systems. Meanwhile, at MIT's Radiation Laboratory – best known, perhaps, for its work on

Figure 12.3 A sideways look at fire control.

radar and long-distance navigation – researchers of the calibre of H. M. James, N. B. Nichols, and R. S. Phillips worked on the further development of design techniques for auto-track radar for AA gun control. And the third institution of seminal importance for fire-control development was Bell Labs, where great names such as Bode, Shannon and Weaver – in collaboration with Wiener and Bigelow at MIT – attacked a number of outstanding problems, including the theory of smoothing and prediction for gun aiming. By the end of the war, most of the techniques of what came to be called classical control had been elaborated in these laboratories, and a whole series of papers and textbooks appeared in the late 1940s presenting this new discipline to the wider engineering community.[9]

Mindell sees the wartime US fire control project as a crucial stage in the development of American national R&D, with implications for the post-war period:

> An understanding of wartime research programs in fire control helps us to understand the NDRC's influence of post-war science policy in America. Section D-2 and Division 7 borrowed the techniques of pre-war private patronage and employed expert project managers with the authority to shape research agendas. After the war, this model was continued by the Office of Naval Research (ONR) and nearly two decades later it was replicated by the Advanced Research Projects Agency (later the Defense Advanced Research Projects Agency, or DARPA). ... Because it was attached to no particular military service, but reported directly to the secretary of defense, DARPA

concentrated on problems, such as computing, that did not obviously fit under any particular service [and] fostered numerous innovations ...[10]

The United Kingdom

David Edgerton has recently argued convincingly for a reappraisal of the history of military technology in Britain, concluding that the contribution of the military to scientific and technological development has been seriously undervalued by historians.[11] Certainly, by the outbreak of the Second World War various military research stations were highly active in such areas as radar and gun laying, and there were also close links between government bodies and industrial companies such as Metropolitan-Vickers, British Thomson-Houston, and others. Nevertheless, in the late 1930s a rather chaotic picture emerges of military R&D in the control field in the UK. The overall system was highly distributed with, for example, components for a particular system being designed at different military research establishments and manufactured by different companies. There was little overall coordination of control R&D, although the Air Defence Research and Development Establishment (later RRRDE, Malvern) had some oversight. Furthermore, as might be expected, there were inter-service rivalries and conflicting project requirements.

A body that contributed significantly to the dissemination of theoretical developments and other research into feedback control systems in the UK was the so called Servo-Panel.[12] Originally established informally in 1942 as the result of an initiative of A. K. Solomon (head of a special radar group at Malvern), it acted rather as a 'learned society' with approximately monthly meetings from May 1942 to August 1945. Table 12.1 indicates the topics discussed at such meetings.

It is interesting to note the collaboration between the US and the UK on wartime R&D reflected in this list: Whiteley reported on US developments early in 1943, and Gordon Brown visited the UK in 1944. Collaboration in this area went back significantly further, however, to Sir Henry Tizard's visit to the US in 1940. Tizard met the NDRC and presented British research on the cavity magnetron, a device of great importance for the development of airborne radar. As part of this visit, he also initiated collaboration between Britain and America on the fire control problem. (There were also significant rivalries!) Ultimately, the Allies were to utilise the US-developed SCR-584 radar and M-9 gun director in defence against German V1 missile attacks on London and Antwerp in 1944. By the end of this period over 50 per cent of V1s were being shot down by the automated system.

The UK wartime infrastructure devoted to servomechanisms did not survive the war, however. Arthur Porter, in a report issued in October 1945, argued strongly for continued government support.[13] He noted that: 'although British scientists and engineers have contributed very appreciably to the theory and design of control systems, America is far ahead in applying this theory in industrial practice'. He attributed this to three factors: '(i) the American industrialist appears to be more "control-minded" than his opposite number [in Britain]; (ii) the number

Table 12.1 A selection of the thirty-six Servo-Panel meetings held between May 1942 and August 1944 (taken from a complete list in Bennett, 1993)

Date	*Presenter(s)*	*Title*
13 May 1942	D. Hartree, A. Porter	The differential analyser and its application to servo-problems
24 July 1942	L. Jofeh	An electrical method of analysis of servomechanisms
8 January 1943	A. H. Whiteley	Some servo developments in the USA
25 June 1943	P. J. Daniell	Interpretation and use of harmonic response diagrams
23 July 1943		Symposium on servo testing
26 November 1943	K. A. Hayes	Use of servos in fire control
24 March 1944	G. S. Brown	Some activities of the servomechanism laboratory, MIT
10 November 1944		Symposium on industrial applications
20 April 1945		Symposium on servo testing
24 August 1945	J. Sudworth	Control systems of the German pilotless missiles V1 and V2

of research personnel engaged on control problems is far greater than the number so engaged in Britain; (iii) the design and application of control systems is included as a subject for postgraduate study at several American universities and technical colleges'. The British government, however, ignored Porter's recommendations and retained the servomechanisms technical committee (which had subsumed the Servo Panel in 1944) as a purely military body, rather than playing the much wider role in both industry and government envisaged by Porter.

Germany

Germany developed successful control systems for civil and military applications both before and during the war[14] (e.g. torpedo and flight control[15]). The period 1938–1941 was particularly important for the development of missile guidance systems. The test and development centre at Peenemünde had been set up in early 1936 and important breakthroughs were made in aerodynamics and propulsion systems during the early years. Work on guidance and control saw the involvement of industry, the government and universities, with companies such as Siemens, Kreiselgeräte and Askania/Möller contributing alongside the technical universities of Darmstadt and Dresden, and the Institute for Oscillation Research in Berlin. An important feature that appears to have aided this success was the concentration of R&D at Peenemünde. According to Michael Neufeld, early contracting out proved not to achieve as much as desired, owing to industry's 'lack of capacity, interest or technological capability [which] forced the rocket program to hire more specialists ...'. The results of centralisation were impressive. Neufeld concludes:

> How could that [technological revolution in rocketry] have been accomplished in five short years? A strong foundation had clearly been laid before 1936 by von Braun's group, and his charisma, intellect and management talent continued to exercise a powerful influence thereafter. University research ... added another critical dimension. The key factors, however, were the investment of massive resources and Dornberger's government-dominated 'everything-under-one-roof' concept for Peenemünde.[16]

However, there does not appear to have been any significant national coordination of R&D in the control field in Germany, and little development of high-performance servos as there was in the US and the UK. Peenemünde was unusual, probably unique, in the tremendous investment by the Reich and enormous support from both the military and the Nazi leadership. Elsewhere, things were different. One of the German pioneers in this field, Winfried Oppelt, remarked in an interview with the author in 1991: 'Everyone kept themselves to themselves ... we didn't know what was being done by the air force, or the navy ... a lot of work was being duplicated, secrecy was a real problem ... links between various research groups could certainly have been better'.[17] Various reasons have been advanced for this overall lack of coordination, for example: early German military success meant there was not the same drive to improve (defensive) fire control; there were complex ideological issues deriving from Nazi views of science and engineering (views that could have both positive and negative influences on technological development, and which have been extensively discussed in German historiography);[18] declining influence of the Army as the war progressed; and the fact that Hitler and Göring rarely sought scientific advice, in contrast to Churchill and Roosevelt.

When we turn to the German situation outside the military context, however, we find a rather remarkable awareness of control and even cybernetic issues.[19] In 1939, the *Verein Deutscher Ingenieure* (VDI), one of the two major German engineers' associations, set up a specialist committee on control engineering. As early as October 1940, the chair of this body, Hermann Schmidt, gave a talk covering control engineering and its relationship with economics, social sciences and cultural aspects – similar to what later became termed 'cybernetics'.[20] Rather remarkably, this committee continued to meet during the war years, and issued a report in 1944 concerning primarily control concepts and terminology, but also considering many of the fundamental issues of the emerging discipline. Figure 12.4, a compilation taken from this report, testifies to an understanding of how general control principles can be applied to a wide range of technological application areas.

Individual engineers in Germany continued to work predominantly in isolation, but it is notable that some from a process control background recognised the generic nature of the emerging discipline. R. Oldenbourg and H. Sartorius,

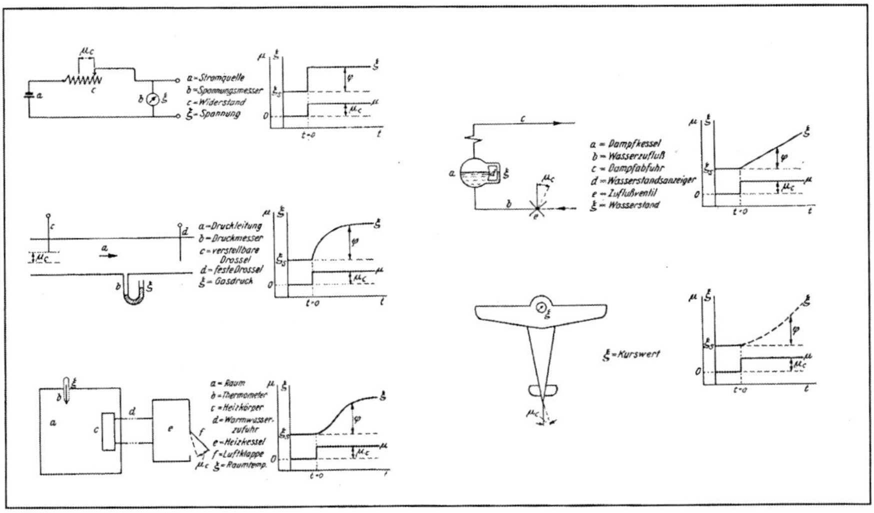

Figure 12.4 A taxonomy of process models from the VDI, 1944.

working at Siemens, were some of the earliest to do this, as their seminal book, published in 1944 and rapidly translated into English, Russian and Japanese, testifies.[21] Although such German researchers did not develop the advanced design techniques characteristic of what came to be known as 'classical control', it is striking how much of this body of theory was developed independently of US and UK work.

The USSR

The Soviet Union saw a great deal of pre-war interest in control, mainly for industrial applications in the context of five-year plans for the Soviet control economy. Developments in the USSR have received little attention in English-language accounts of the history of the discipline apart from a few isolated papers.[22,23,24] It is noteworthy that the *Kommissiya Telemekhaniki i Avtomatiki* (KTA) was founded in 1934, and the *Institut Avtomatiki i Telemekhaniki* (IAT) in 1939 (both under the auspices of the Soviet Academy of Sciences, which controlled scientific research through its network of institutes). The KTA corresponded with numerous Western manufacturers of control equipment in the mid-1930s and translated a number of articles from Western journals. Indeed, Russian researchers in the field demonstrated an awareness of many of the developments of the 1930s in the UK and the USA. These were reported at a 1940 Moscow conference, probably the first it its field. The early days of the IAT work were marred, however, by the 'Shchipanov affair', a classic Soviet attack on an IAT researcher for so-called 'pseudo-science', which detracted from technical work for a considerable period of time.

The other major Russian centre of research related to control theory in the 1930s and 1940s (if not for practical applications) was the University of Gorkii (now Nizhnii Novgorod), where Aleksandr Andronov and colleagues had established a centre for the study of non-linear dynamics during the 1930s.[25] (This approach to dynamics was instrumental in the next generation of control theory, so-called 'state space' techniques, from the mid-1950s onwards in the West.) Andronov was in regular contact with Moscow during the 1940s, and presented the emerging control theory there – both the non-linear research at Gorkii and developments in the UK and the USA. Nevertheless, there appears to have been no coordinated work on control engineering in the USSR, and the IAT in Moscow was evacuated when the capital came under threat from invading German forces. However, there does seem to have been an emerging control community in Moscow, Nizhnii Novgorod and Leningrad, and Russian workers were extremely well informed about the open literature in the West.

Aftermath

The immediate post-war years saw the appearance of numerous specialised university courses, textbooks, and translations of textbooks, not to mention much military R&D in automatic missile guidance as the Cold War intensified. Table 12.2 indicates the transfer of Western classical control engineering ideas to the Soviet Union, but soon afterwards Soviet achievements in the field were also translated into English. Cover-to-cover translation of Russian journals in a range of scientific fields began in the 1950s.

Professional engineering bodies in various countries soon took an interest in the new discipline, setting up new specialist sections or expanding the few that had existed before the war. Conferences, too, abounded. One of the first important conferences was held in London, in 1947, organised by the Institution of Electrical Engineers (IEE), which drew together mainly UK researchers. Probably the first international conference on control was held at Cranfield, UK, in 1951, organised by the UK governmental Department of Science & Industrial Research together with the Institutions of Electrical and Mechanical Engineers (IEE and IMechE). Although most participants were from the US and the UK, there was also a significant German participation. Five years later, an international confer-

Table 12.2 Early translations of control texts into Russian

Author	*Date of original publication*	*Russian translation*
Oldenbourg & Sartorius	1944	1949
Bode	1945	1948
McColl	1945	1947
James et al.	1947	1951
Lauer et al.	1947	1948

ence in Heidelberg, Germany, attracted significant Soviet participation (six full papers) as well as forming the occasion of an agreement to form the International Federation of Automatic Control (IFAC). The Heidelberg conference is also seen by some as marking a turning point in the development of control theory: alongside the classical control that emerged from the war years, the new 'state space' or 'modern' control theory, based on a very different approach to system dynamics, began to be presented.[26]

Conclusion

Even this very cursory account demonstrates that there were important developments in control engineering during the Second World War in various warring nations, not simply in the US and the UK. Support for such R&D varied greatly from country to country, and reflects the ideologies and traditions of the various nations. On the one hand, there were hugely resourced projects such as the fire control project in the US (and to a lesser extent the UK) and the missile project in Germany. Whereas the former led to a generic approach to control, and the development of a range of design techniques for closed-loop systems, the latter appears to have remained highly specialised, and not to have led to developments outside missile guidance. At the same time, the Servo Panel in the UK provided an effective, if fairly informal, means of disseminating the new ideas in the field, including those from America. Meanwhile, the VDI in Germany was quietly meeting and developing independently of the Allied work an understanding of the general systems nature of control engineering. In many ways, those Russian scientists and engineers interested in control problems were taking a different route, developing an approach to non-linear systems that came to the fore in the post-war period. Yet the Russians were well aware of the non-classified Western work, and in a good position to take control engineering further during the Cold War.

Notes

1 O. Mayr, *The Origins of Feedback Control* (Cambridge, MA: MIT Press, 1970).
2 S. Bennett, *A History of Control Engineering 1800–1930* (Stevenage: Peter Peregrinus, 1979).
3 S. Bennett, *A History of Control Engineering 1930–1955* (Stevenage: Peter Peregrinus, 1993).
4 D. Mindell, *Between Human and Machine* (Baltimore: Johns Hopkins University Press, 2002).
5 Mindell 2002, 175.
6 K-H. Fasol (ed.), 'Great names and the early days of control in Germany', *Automatisierungstechnik*, 9 (2006), 462–73
7 Mindell 2002; Bennett 1993.
8 See Mindell 2002, Appendix B for a full list
9 C. C. Bissell, 'Textbooks and subtexts', *IEEE Control Systems Magazine*, 2 (1996), 71–8.
10 Mindell 2002, 311.

11 D. Edgerton, *Warfare State. Britain, 1920 – 1970* (Cambridge: Cambridge University Press, 2006).
12 A. Porter, 'The Servo-Panel – a unique contribution to control-systems engineering', *Electronics and Power*, October (1965), 330–3.
13 A. Porter, *Memorandum on 'The Application of Control Systems in Industry'*, (London: Ministry of Supply, October 1945).
14 B. R. Kröner et al. (eds), *Organisation und Mobilisierung des Deutschen Machtbereichs. 2. Halbband. Kriegsverwaltung, Wirtschaft und Personelle Ressourcen 1942–1944/5* (Stuttgart: Deutche Verlags-Anstalt, 1999).
15 M. J. Neufeld, *The Rocket and the Reich* (New York: The Free Press, 1995).
16 Neufeld 1995, 108.
17 C. C. Bissell, 'Six decades in control: an interview with Winfried Oppelt', *IEE Review*, January (1992), 17–21.
18 Kröner 1999, 659–92.
19 K-H. Fasol, 'Hermann Schmidt: pioneer in control and cybernetics', *IEEE Control Systems Magazine*, April (2002), 24–8.
20 H. Schmidt, 'Regelungstechnik – die technische Aufgabe und ihre wissenschaftliche, sozialpolitische und kulturpolitische Auswirkung, *Zeitschrift VDI*, 4 (1941), 81–8.
21 R. C. Oldenbourg and H. Sartorius, *Dynamik selbsttätiger Regelungen* (Munich: Oldenbourg Verlag, 1944).
22 C. C. Bissell, 'A. A. Andronov and the development of Soviet control engineering', *IEEE Control Systems Magazine*, 1 (1998), 56–62.
23 C. C. Bissell, 'Control Engineering in the former USSR: some ideological aspects of the early years', *IEEE Control Systems Magazine*, 1 (1999), 111–17.
24 C. C. Bissell, 'The "First All-Union Conference on Automatic Control", Moscow, December 1940', *IEEE Control Systems Magazine*, 1 (2002), 15–21.
25 A. Dahan Dalmedico, 'Early developments of non-linear science in Soviet Russia: The Andronov School at Gorky', *Science in Context*, 1/2 (2004), 235–65.
26 Bennett 1993, 203.

13 British cryptanalysis

The breaking of 'Fish' traffic

J. V. Field

The use of radio communications in warfare makes cryptography important. Anyone can receive the messages, but the sender wishes them to be understood only by the intended recipient. The importance of cryptography accordingly became apparent in World War I, and the British Government Code and Cypher School (GC&CS) was set up in 1919.[1] In 1939, it was transferred to Bletchley Park, an undistinguished, moderately-sized Victorian country house conveniently situated for rail access from London, Oxford and Cambridge.

All the work was, of course, highly secret, but the choice of location suggests that the staff were not intended to work in complete isolation from the milieux from which most of the senior members were recruited. The gardens of the house were filled with rows of huts, so that from the air the establishment seemed to be an ordinary army camp.

Radio also makes it highly desirable – and thus in the longer run inevitable – that the processes of encryption and decryption of signals will use machines. When a radio operator can tap out a message in minutes it obviously makes no sense for a cipher clerk to require several hours to work out what should be sent or what message has been received. The encryption and decryption machines that will concern us here are of two kinds. The first is the one the Germans called Enigma, introduced in the 1920s. Similar, though not identical, machines were used by the British and the Americans; the British one was called 'Typex', the American one 'Sigaba'. (Sigaba was considerably more complicated than the other two, but all three employed the same basic principles.) Machines that worked on a different principle were used for encrypting German teleprinter communications. It is the breaking of traffic from two machines of this second kind that is the main subject of the present paper.

The business of the staff at Bletchley Park was signals intelligence, specifically to read the encrypted radio messages picked up by a series of interception stations. There is, of course, no solid evidence for the assertion, sometimes made in histories of the work of Bletchley Park, that the successful reading of enemy traffic shortened the war. However, it appears from the historical record that the British Government eventually decided that the intelligence gathered from reading German signals was valuable enough to warrant assigning considerable resources to the cryptanalysis. When, on 21 October 1941, four senior members of the staff of Huts 6 and 8 at Bletchley Park ignored the official channels of communication and sent a letter directly to the Prime Minister asking for additional funding, they got it.[2] That funding was in connection with work on breaking Enigma traffic.

The work breaking the ciphers used in teleprinter messages, called 'Fish' by the cryptanalysts – the work that forms the main subject of the present paper – seems to have benefited from the example of the success of the breaking of Enigma. The teleprinter traffic carried very high-level communications, between generals in the highest military headquarters and those in command of field operations (occasionally from Hitler to a general), and reading it therefore yielded strategic rather than merely tactical information. This doubtless helped to ensure that, as the cryptanalysis proceeded, resources were found for the construction of a number of electrical machines of unprecedented complexity. Some of these machines are obviously part of the ancestry of the modern electronic computer, but whether one wishes to regard any of them as actually being a computer is essentially a matter of how one defines the term 'computer'. None of the Bletchley Park machines used a stored program in the present-day sense – the first machine to do so was the Manchester 'Baby' of 1948 – and none of them used 'branching', that is the machines did not take decisions between several possible logical paths.

The work at Bletchley Park seems also to have played a part in establishing an initial connection between computing (and what is now called 'computer science') and mathematics, because, as we shall see, there was an increasing use of mathematicians as cryptanalysts. Thus, it was one of the effects of this wartime work that mathematics began to look a less rarefied academic discipline than before – but the later history of that change belongs to a much larger story that cannot be discussed here.

Enigma

In the early part of the war, most of the traffic intercepted and sent to Bletchley Park for analysis – it was carried by despatch riders on motorcycles – came from the Enigma machine. The signals were in Morse code, which uses a series of dots and dashes (shorter and longer signals) to indicate letters of the alphabet and punctuation marks to spell out messages letter by letter. The overall design of the machines that produced the transmissions was well known to the British cryptanalysts, but as the war continued some variants were introduced. In the end, the difficulties they presented for the cryptanalysts did not prove to be insuperable, though there were periods during which no traffic could be read.

A method for attacking the Enigma cipher had already been devised, before the war, by a team of Polish mathematicians led by Marian Rejewski (1905–1980). The Polish cryptanalysts passed their work on to the French and the British, so that by the time hostilities broke out, and the German army rapidly advanced to positions which made it necessary to use radio rather than landlines to carry its communications, the British were already well placed for reading German Enigma traffic. And by the spring of 1940 read it they did.[3] There were setbacks when the Germans increased the number of wheels used in their

Enigma machines, but eventually the workers at Bletchley Park were able to mechanise the breaking of Enigma messages. The electromechanical machines that found the settings of the wheels used in an encrypted transmission were called 'bombes', after the name 'bomba' (plural 'bomby') that the Polish group had given to the similar but simpler apparatus used in their cryptanalysis. As their co-designer Alan Turing (1912–1954) put it, the bombes worked not by finding the correct solution but by eliminating those that were impossible. Turing's account of work on Enigma, known at Bletchley Park as 'The Prof's book' but properly called *A treatise on Enigma*, is now held by the National Archives and Records Administration (NARA) in Washington DC.[4] Once the wheel settings had been found, a suitably set British Typex machine could be used to produce decrypted text. We may note, however, that as one would need to type in the encrypted message, that is gibberish, a skilful typist was required. Bletchley Park employed a large number of women, Wrens (Women's Royal Naval Service) and ATS (Auxiliary Territorial Service) girls[5] to operate machines and carry out typing and other routine or administrative tasks.

The bombes were essentially a bank of Enigma simulators. It was thought that details of their design had been lost, but a set of engineering drawings was found to have been preserved at Bletchley Park's successor institution, the Government Communications Headquarters (GCHQ, situated outside Cheltenham), and it has been possible to rebuild a bombe. The bombe used for attacking a three-rotor Enigma machine was a substantial structure, the parts being assembled on a steel frame with base 2.21 m by 0.8 m, and height 1.78 m, mounted on heavy-duty castors. There were three sections, one above the other, each containing twelve columns of three wheels each. The three sections worked independently, each on a separate set of solutions, but if one stopped, on a possible solution, then all stopped. The wheels and a multitude of connection rods were driven by a direct-current (DC) electric motor, connected to the mains through a mercury-arc rectifier. The machines were manufactured in Letchworth (Hertfordshire), by the British Tabulating Machine Company (BTM, later to become ICL).

Rebuilding a bombe has shown that a significant effort must have gone into the original machines. Only 15 per cent of the parts were standard items for BTM. It has become clear that the machines represented not only a considerable intellectual achievement on the part of the designers (Alan Turing and Gordon Welchman of Bletchley Park and Harold Keen of BTM) but also an important commitment of resources for manufacture.[6]

For reasons of security, work at Bletchley Park was divided between separate groups, whose members were strictly forbidden to discuss their work with anyone other than their immediate colleagues. However, a wider view was available from the higher levels in the organisation; the successful mechanisation of the breaking of Enigma traffic made it seem reasonable to hope that machines could also be used against other ciphers. And as the project progressed, staff were transferred from working on Enigma to work on other ciphers.

'Fish' Traffic

Messages encrypted in Enigma were received in Morse code. In January 1941, the British interception stations began to receive signals transmitted in the twittery tone used for teleprinters.[7]

The various links over which these transmissions were made were given the names of fish: 'Whiting', 'Bream', 'Gurnard' and suchlike, including (presumably humorously) 'Jellyfish'. The ciphers used took on the names of the links. It turned out that there were three distinct varieties of cipher machine, which were given the names of the corresponding links: Thrasher, Tunny and Sturgeon.[8] The one that will concern us here is Tunny, a machine manufactured by Lorenz.[9] Whereas Enigma ciphers were used for traffic at field-command level, it turned out that the teleprinter traffic encrypted in Fish ciphers carried high-level communications between generals and between Hitler and his generals. These high-level ciphers were much more secure, that is much more difficult to break, than Enigma. It did, however, prove possible to break them, though even by 1945 not all messages could be read and one of the cryptanalytic team said (in 2000), 'We were hanging on by our fingernails; if the Germans had changed two things at once instead of only one at a time they would have lost us'.[10] At the end of the war, between May and September 1945, three of the cryptanalysts who had worked on these teleprinter ciphers, I. J. (Jack) Good (b. 1916), Donald Michie (1923–2007) and Geoffrey Timms (1903–1982), put together a detailed report on their work, entitled *General Report on Tunny with Emphasis on Statistical Methods*. This report, about five hundred typed pages, was declassified and released to the Public Record Office in September 2000 (press marks HW 25/4 and HW 25/5). It has provided material for much of what follows. The Report was written by a number of participants in the work, sometimes directly and sometimes by the editors making use of pre-existing 'Research Logs' that recorded specific results.[11] No authors are named for any part of the Report; even the names of the editors do not appear in the front matter of the Report (which has no title page) and the Research Logs seem to have been lost, so it is an intricate task for future historians to work out who wrote what. However, the overall authority of the account itself is not in doubt. One should bear in mind that the authors were perhaps not always immediately involved in the work they describe, but they were in a position to ask the principals about what were then relatively recent events.

The cryptanalysts of course knew that teleprinters used five-hole punched tape which employed Baudot code to indicate letters and symbols.[12] (The operator could either type directly into the machine using the typewriter keyboard or could feed in a pre-prepared punched tape. Output was as letters typed on paper or as a punched tape.) However, in other respects the cryptanalysis started from zero: the encryption machine was unknown. In fact, it seems to have been relatively unfamiliar to the German operators also: the first, crucially important, break was made because of errors by a cipher clerk. On 30 August 1941, by which time the cryptanalysts already knew something of Tunny practice, a clerk sent the same message twice, without altering the wheel settings (which meant that the key was exactly the same) but making many small changes in what was typed, for instance in the use of abbreviations, which produced small offsets in the relation between plain-text and key. Once noticed and identified, both these errors could be exploited. The chief military cryptanalyst at Bletchley Park, John Tiltman (1894–1982), exploited them with spectacular success: he read the message (that is he found the original German plaintext) and he isolated a stretch of nearly four thousand characters of pure key. Interpreting this was not easy, but in early 1942 the young Cambridge mathematician William Tutte (1917–2002) succeeded in deducing the structure of the machine that produced the key.[13]

The technical details of the machine are of some significance here because they explain the choice of methods of attack used by the cryptanalysts. The five 'channels' of the Baudot code on the tape were encrypted separately. Each channel consisted of a stream of 'bits' each of which was either a 0 or a 1, that is a blank or a hole in the tape, which the cryptanalysts signified by a dot (for 0) or a cross (for 1).[14] To this stream the Tunny machine added the key stream, a stream of 0s and 1s of its own devising. This key stream was generated by twelve wheels carrying elements that could be adjusted (by hand) to give either 0 or 1. The wheels were of three kinds, which the cryptanalysts called by the Greek letters chi, psi and mu. The five chi wheels turned on one place for every new bit, the five psi wheels might or might not turn depending on a complicated set of conditions involving the mu wheels, the previous chi and psi bits, the twice-previous plain text bits and other conditions that could be set by the operator.

The numbers of adjustable elements on the wheels were as follows:

chi wheels: 41, 31, 29, 26, 23;

psi wheels: 43, 47, 51, 53, 59;

mu wheels: 61, 37.

The numbers are all prime, except for the 26 on one of the chi wheels, but as there is no 13, the numbers are all mutually prime, which is all that is required to maximise the length of the cycle (that is the number of bits before the key pattern begins to repeat itself – ignoring irregularities in the motion of the psi wheels, which are intended to disrupt the cycle). The addition of the key stream does not permit carrying (that is one channel cannot affect another) so we have addition modulo 2, in which:

$$0 + 0 = 0$$

$$0 + 1 = 1$$

$$1 + 0 = 1$$

$$1 + 1 = 0.$$

(Readers of a mathematical bent will observe that this makes subtraction the same as addition. So adding the key again is the same as subtracting it and the cipher apparatus attached to the teleprinter can thus treat incoming and outgoing messages in exactly the same way.) There was one other form of adjustment: the initial starting positions of the wheels could be varied.[15]

Thus, to read a message the cryptanalysts needed to determine the pattern of 0s and 1s round each wheel (to 'break' the wheels) and to find their starting positions (to 'set' the wheels). This is in some respects similar to what was required for reading Enigma messages, but the form of encryption in the Tunny and Enigma ciphers is radically different. Whereas Enigma deals with characters, for instance turning an A into a Q, Tunny operates not on characters but on bits. The ciphers used during the period of the Cold War were mainly of this bit-oriented type, which may go some way to explaining why the detailed report by Good, Michie and Timms remained classified for so long. The use of part of the plaintext to construct key, called 'autoclave' in the discussion of the psi wheels in the *General Report*, was not new at the time. It is now known as 'autokey'. The rather simple form used in the Tunny machines seems to have given the cryptanalysts little trouble,[16] but more elaborate autokey systems were still in use in 2007.

Mr Newman's section

A Research Section had been set up in August 1941, reporting to the head of the military section, that is to Colonel Tiltman. In July 1942 some of its functions in regard to Tunny were taken over by a new section, run by Major Ralph Tester (1901–1996). Its purpose was to apply the methods of attacking Tunny that had been developed by Tiltman and Tutte. In the same month, July 1942, current Tunny traffic was read for the first time. The style of nomenclature used at Bletchley Park owes much to British Public School slang – a characteristic that is indicative of the social origins of most of the senior staff – so the group run by Major Tester became known as 'The Testery'.

Max Newman (1897–1984) was born in London, the son of a German immigrant (the family name had been Neumann) and an English mother. By the 1930s he was a successful mathematician, held in high regard by his professional colleagues. He was elected a Fellow of the Royal Society in 1939. At the outbreak of war he seems at first to have supposed that his German ancestry might make him ineligible for war work, but that turned out not to be so and in 1942 he temporarily laid aside his lectureship in mathematics at Cambridge University, and his fellowship of St John's College, to work at Bletchley Park, where he took up his appointment on 31 August.[17] He was employed in working on Tunny in the Research Section and, having an interest in automating mathematical and logical tasks, thought it would be possible to use machines for this one.[18]

Here it is again relevant that we are considering not a general mathematical or logical task but specifically that of cryptanalysis: speed is important. The sooner one reads intercepted messages the more useful they are. The complicated nature of the Tunny machine made the cryptanalysis laborious, and – presumably thanks to the alertness of the German signals staff – procedures had been altered so that the traffic no longer supplied 'depths', that is the kind of multiple or repetitive transmission that Tiltman and Tutte had exploited in making their crucial breaks and consequent 'diagnosis' of the structure of the machine. Alan Turing, whose position was that of a general consultant, joined in the Testery's work on Tunny and invented a method of finding the key stream by using differencing – that is subtracting successive bits one from another, thus finding the change from one bit to the next on the tape – but his method (known as 'Turingery') was slow. In November 1942 Tutte (in the Research Section) devised a statistical approach. It did not rely on depths and it could be used on a single message (provided it was long enough). The method considered pairs of settings of the first two chi wheels;[19] and it involved differencing and counting of characters, all repeated for different possible cases. The method was powerful but slow. Moreover, the cryptanalysts were looking for small effects, so the counting had to be done accurately. Omitting a character can, as it were, push the cycles out of phase, making the pattern more difficult to see. As the cryptanalysts put it in their Report:

> The standard of accuracy needed before there was any possibility of success was very much higher than would ordinarily be required of this kind of apparatus, or of operators. A single letter omitted in a tape destroyed the value of the run and the ordinary length of a tape was about 3000 letters.[20]

This was, of course, written when machines were in operation, long after the decision to mechanise had been taken. However, it carries echoes of why, a hundred years earlier, Charles Babbage (1792–1871) thought that machines should be used for calculating and printing mathematical tables. The idea of attempting a statistical attack on Tunny and that of using machines seem natural bedfellows. In November 1942 Newman suggested the use of electronic counting machines.[21] Such machines had been invented in Cambridge before the war, so Newman was in a good position to know of their existence and capabilities.

In December 1942 Newman was allowed to set up a new section, with the specific remit of mechanising the breaking of Tunny. This group was known as 'The Newmanry'.[22] At first there were only two cryptographers: Newman himself and Donald Michie. Michie had left Rugby School in 1942, with a scholarship in Classics to Balliol College, Oxford, and (intending to fill in time before receiving call-up papers) had attended one of the cryptography courses that Tiltman had set up in Bedford, which were of course intended to produce potential staff for GC&CS.[23] The next recruit was Jack Good, a Cambridge mathematician who had been working on Enigma, in Hut 8, since March 1941.[24] Probability theory was not an established research interest at Cambridge at this time: in 1936, when Alan Turing wrote his Smith's Prize essay on a Probability topic, the university's readers did not recognise that the theorem he proved had in fact been known for several years.[25] However, Good had taken an interest in Probability theory since his schooldays[26] and an examination of the first part of the *General Report* shows that much of the statistical work he carried out in the course of the attack on Tunny was later to find its place in his highly influential book *Probability and the Weighing of Evidence* (London: Charles Griffin, 1950), though of course with no hint of its cryptanalytic origins. Turing, who (as we have seen) had done some work on Tunny in the Testery, also participated in the activities of Newman's group, and seems to have persisted in his habit of proving theorems for himself rather than finding them in the works of others: Good had to tell him that one result was Bayes' Theorem (named after the mathematician who had first proved it in the eighteenth century).[27]

The Machines

Newman's idea of using electronic counting machines became reality: a suitable machine was designed, largely by the physicist and electrical engineer C. E. Wynn-Williams (1903–1979) of the Telecommunications Research Establishment (TRE).[28] The date of the order for the machine is given in the *General Report* as January 1943. The machine arrived at Bletchley Park in June and acquired the name 'Heath Robinson' after the complicated and fantastical machines depicted by the well-known cartoonist Heath Robinson (1872–1944). It seems that the names of machines were generally invented by the women (WRNS) who operated them. In this case, the British habit of using only surnames led to the name becoming 'Robinson'. Later models were called Old Robinson and Super Rob.

An eight-page, month-by-month 'Chronology', running from June 1941 to June 1945, is provided as section 74 of the *General Report*.[29] This takes the form of a table with five columns. The first gives the year and month, the second is headed 'Changes in Tunny' and its first entry is 'SZ40 first Tunny link'[30] for June 1941, the third column is headed 'Organisation changes' and its first entry is 'Work in Research Section starts' for June 1941, the fourth column is headed 'Machines' and its first entry is 'Decoding machine ordered' for April 1942, the fifth column is headed 'Theoretical Discoveries and Achievements' and its first entry is 'Machine broken for Aug 1941' for February 1942. This entry refers to the completion of the process of understanding the overall design of the Tunny machine, plus wheel patterns, from the work of Tiltman and Tutte on the message received in August 1941 (discussed above). The next break of a machine, that is the finding of wheel patterns, is recorded in April, and was for the traffic of the previous month.[31] The 'decoding machine' referred to in the entry for April 1942 is a Tunny machine simulator, designed to decrypt the traffic when the correct wheel patterns and wheel settings are known. The machine was delivered in June. As we have already noted, current Tunny traffic was read for the first time in the following month, July 1942, marking the beginning of the work of the Testery. After that, the machines considered worthy of mention in the Chronology are those for the Newmanry.

The Robinson machines (by the end of hostilities there were three in operation) are designed to carry out some of the more laborious procedures involved in the breaking of Tunny traffic. The *General Report* categorises Robinsons as 'counting and stepping machines' and their action is described as follows:

> These machines are given two teleprinter patterns, combine them in some way and count the number of places of the combined pattern in which a certain condition is satisfied.
>
> An essential feature is that these counts must be made with the two patterns in all possible relative positions i.e. one pattern must 'step'.

An example is given and we are then told that 'at each setting the answer is, of course, a number'.[32] The style of the *General Report* marks it as written by specialists for their peers and their successors, so the simplicity of this last statement is a reminder that this kind of machine was not a familiar sight in the early 1940s.

The *General* Report does sometimes consider matters historically – for instance in providing the chronological table to which we have referred – but the Robinsons are treated rather condescendingly. The subsection devoted to them in a section entitled 'Machines' in the introductory part of the *General Report* begins with a reference to the machine that was developed from the Robinson, the Colossus:

> In pre-Colossus days the old Robinson did much of the work now assigned to Colossus, and, considering its primitive character, did so with remarkable success.[33]

The design of the Robinsons allowed their operation to be flexible. This was necessary because the cryptanalysts knew from experience that their task might change if the Germans decided to modify the Tunny cipher. The design of successive Robinsons was also modified to take account of difficulties encountered in operating each model. Indeed the reasons for these changes bulk large in the fuller treatment of the Robinson machines in the sections of the *General Report* specifically dedicated to them, 52 ('Development of Robinson and Colossus') and 54 ('Robinson').[34] Here again, the Robinson is subordinated to the Colossus, in section 52 by the Robinson being described almost entirely in terms of the features that proved inconvenient and led to the development of later models and eventually to Colossus, and in section 54 simply by the fact that the Colossus has been treated at much greater length in the previous section.[35] The difficulties with the Robinson are largely practical ones, to do with the hardware. Nine 'handicaps' are listed, for example that the machine did not print the numbers it found, which had to be written down in haste by the operators, who of course were liable to make errors.[36] Details of hardware are precisely what will not be supplied in the *General Report*; section 51, entitled 'Introductory', begins, under the heading 'Character of chapters 51 to 58':

> This is a strictly functional and non-technical account of the machines used. A technical section is to be prepared by the post office engineers.
>
> Some attempt is made to avoid statements technically false, but none to avoid statements technically vague.[37]

The report by the Post Office engineers has not been found. We are thus left with a rather sketchy account of the passage from Robinson to Colossus. Nevertheless, it seems fairly clear that the details of the development of the new design largely concern hardware, though there seems also to have been an increasing sense that the 'useful' Robinsons indicated that it should be possible to make something that worked faster. The principles of the Robinson were, apparently, regarded as satisfactory, though the *General Report* says so in a deterministic style that presumably owes something to hindsight from experience with Colossus:

> In the experimental stages of Tunny-breaking, though other forms of machine were considered, it was inevitable that one using Robinson principles should be chosen because (α) it is easy to make; (β) it can be adapted to any wheel length by preparing suitable tapes.[38]

We are then given a list of nine 'handicaps' of the early Robinson design that, with one exception, were remedied in the later models, Old Robinson and Super Robinson. The handicaps that were remedied include the non-printing of results already referred to and an inability to deal with tapes of less than 2000 characters. The handicap that is mentioned as not having been overcome is the inability to carry out 'spanning' that is to process only a set part of the tape, a procedure 'whose value was overlooked till later'.[39] These improvements were all incorporated in the first designs for Colossus. Storage of data became a feature of this new design. Tapes needed to be read repeatedly, so it was clear that speed could be improved if some way were found to store information from at least one of the tapes being compared. This would require the use of many more valves.

The authorities who had the power to approve the project for a new machine developed from the Robinson responded slowly – perhaps understandably since there was doubt whether a machine containing so many valves could be expected to be reliable – and they seem in the end to have been circumvented by the Post Office research laboratory going it alone. Staff at the laboratory, at Dollis Hill, in north-west London, about 40 miles (65 km) from Bletchley Park, had helped TRE with the design and construction of the Robinsons. In February 1943 – apparently without official participation by staff from Bletchley Park – a Post Office team led by Thomas ('Tommy') Flowers (1905–1998) set to work on the design of the new machine.[40] The start of the Post Office's work is not recorded in the Chronology in the *General Report*, though, as we have seen, dates are given for the ordering of Robinsons. Nor does the Chronology record that the new machine was demonstrated successfully in an eight-hour run in December 1943. That demonstration proved the thyratron valves were sufficiently reliable. (It had turned out that valves failed much less often if the current was never switched off – which was how Colossus was used.) In regard to the design and development processes for the new machine, the Chronology records only that the first operational Colossus machine was delivered to Bletchley Park in February 1944. The Chronology also records that in the previous month a teleprinter line had been installed linking the interception station, at Knockholt, in Kent, with Block F at Bletchley Park. The installation of this direct line suggests that the authorities had at least accepted that Newman's group were there to stay in breaking Tunny traffic. Because of its huge size, the operators (WRNS) called the new machine Colossus. The machine consisted of twelve racks of apparatus, each about 2.5 m high and, allowing for circulation space for operators and engineers to move between the components, took up most of the floor space of a room measuring about seven metres square. (A Robinson consisted of a single rack, a 'bedstead' base about 1 m by 3 m and height 2.5 m, plus a small typing machine.)

Moves towards designing and building Colossus started very soon after the construction of the first Robinson. The Chronology in the *General Report* dates the Robinson to January 1943 and work began on Colossus in February. As, again according to the Chronology, Newman first put forward his suggestion for an electronic counting machine only in the previous November, it hardly seems possible to regard the Colossus as anything other than a developed version of the Robinson; experience with using Robinsons must have influenced development work on the later machine. In any case, the intellectual thrust was already made with the Robinsons. Those machines already took the step of automating a mathematical task; that is an ordered series of operations, rather than merely a single mathematical procedure such as subtraction. In this sense the Colossus represented a practical rather than an intellectual development, simply using different electrical components to give a better performance. It was this better performance that ensured that Colossus took precedence over Robinson in the *General Report* as it eventually did in the work the authors are describing: by the end of the war there were three Robinsons in operation and ten Colossi.

We are concerned here not with a research report but a report on a practical project in which all research was task-oriented. With hindsight we see intellectual achievements, but this was war work and the 1945 Report duly presents it as such, though (as we have seen) the final column of the chronological table does allow the recording of 'theoretical discoveries' as well as 'achievements'. In fact, Newman allocated his staff time for theoretical research, as formal 'research shifts', and the ethos seems to have been rather close to that of an academic department in a university. All cryptographers worked on all tasks, and any one of them was permitted to call a meeting if an issue arose that seemed to merit general discussion. Such a meeting was called a 'Tea party', though no tea was served.

The tolerant collegial attitude had its limits, however. The improved performance of Colossus compared with Robinson apparently made it more easily adaptable to carrying out different procedures. This had a practical purpose, because changes were repeatedly made in the Tunny cipher and there always remained some traffic that was not broken, on which new techniques might be tried. Newman obviously recognised that Colossus came closer than any other machine had to being the 'universal machine' Turing had described in 1936 in his paper 'On computable numbers'.[41] And he realised his colleagues would have noticed the same. Over sixty years later Donald Michie still remembered the sternness with which he was told it was forbidden to play with Colossus.

The unprecedented speed of the machine undoubtedly exercised a fascination. In the rather dry *General Report* there are some passages that give a sense of intellectual drive, but rather few that give a sense of exhilaration. One such is the passage that captures the excitement of seeing an automatic machine at work, moving fast through calculations, in an unfamiliar kind of automation. The description is headed 'Impression of Colossus'. It reads:

> It is regretted that it is not possible to give an adequate idea of the fascination of a Colossus at work: its sheer bulk and apparent complexity; the fantastic speed of thin paper tape round the glittering pulleys; the childish pleasure of not-not, span, print main heading and other gadgets; the wizardry of purely mechanical decoding letter by letter (one novice thought she was being hoaxed); the uncanny action of the typewriter in printing the correct scores without and beyond human aid; the stepping of display; periods of eager expectation culminating in the sudden appearance of the longed-for score; and the strange rhythms characterizing every type of run: the stately break-in, the erratic short run, the regularity of wheel-breaking, the stolid rectangle interrupted by the wild leaps of the carriage-return, the frantic chatter of a motor run, even the ludicrous frenzy of hosts of bogus scores.
>
> Perhaps some Tunny-breaking poet could do justice to this theme; but although an ode to Colossus and various fragments appeared, all seemed to have been composed in times of distress and despondency, and consist almost wholly of imprecation or commination.[42]

The famous, and historically important, machines are the large ones, but right up until the end of the war there were also many smaller machines in use (and at least two being developed), each designed for a specialised task, such as the Dragon used to set a 'crib' (a piece of text believed to occur within the plaintext of a given message) in all possible positions against a tape of ciphertext from which the chi-wheel component of the key had been removed, a procedure that allowed the psi-wheel component to be identified. The operation was called 'dragging' the crib, hence the name of the machine, which was used in the Testery.[43] The list of smaller machines in the *General Report* is almost comically complete, as if it were intended as an inventory; it includes many entirely commonplace items, such as the slide rule, which attracts the acid comment 'Many of the slide rules used lack logarithms and have elaborate useless scales'.[44]

The *General Report* provides a table showing which machines were in use in various places in May 1943 and in May 1945. In May 1943 there was one Robinson, one Tunny machine and one machine to help in sticking tapes together to make a loop to run on Robinson (or, later, on Colossus). By the end of the war there were many more machines. In May 1945 the Newmanry had two Robinsons and ten Colossi; the Testery had three smaller machines (two Dragons and a machine called Aquarius, described as 'on test') plus thirteen Tunny machines (for decoding), while the Newmanry had four Tunny machines used for cryptanalysis,[45] plus thirty-three minor items.[46]

It is thus clear that mechanisation did not take place only in Newman's group. Smaller machines were used in the Testery, where hand methods were employed for wheel setting and for removing the effects of the psi wheels from transmissions from which the Newmanry had removed the effects of chi wheels. Although it was recognised that, in principle, the whole task of breaking the cipher could be mechanised, allowing a Tunny machine to be set up so that messages could be decrypted, there seems to have been no attempt to actually carry out such a programme. Perhaps there would have been if the war had gone on longer. At least two of the ten operational Colossi were transferred to GCHQ after the war. Perhaps complete automation of cryptanalysis was put into operation there.

Staff

The increase in the number of machines went with an even steeper increase in the number of staff employed in Newman's group. The numbers are given for six-monthly intervals, starting in April 1943, when there were 2 cryptographers and 16 Wrens (WRNS), making a total of 18. In April 1945 there were 2 administrators, 22 cryptographers, 28 engineers (15 for maintenance, 13 for construction) and 273 Wrens, making a total of 325.[47] We may note that the Wrens, employed as ancillaries, operating machines and carrying out routine tasks such as calculations, thus formed a roughly constant proportion of the staff.

Interestingly, the *General Report* also presents a discussion of the backgrounds, not to say the characters, of the Newmanry cryptographers. The text begins:

> The first thirteen men to join the Section as cryptographers were drawn from other sections of GC&CS. In experience and infectious enthusiasm they preserved their lead to the end, and there were few in the section not affected by their keenness. After July 1944 they were joined by men from other war jobs and men straight from the universities.[48]

This is victors' history, written in the moment of victory, but it was objectively true that Newman's group had done what it was set up to do. Of the three authors of the *General Report*, the first two, Jack Good and Donald Michie, belonged to the initial thirteen. The third author, Geoffrey Timms, reached Bletchley Park in September 1944, at the age of 41, after a university career in mathematics.[49] All members of the Newmanry who have written about their experiences agree that Newman was an exceedingly able and inspiring leader.

The table introduced by the passage just quoted analyses the two groups of recruits in four ways. The first is by their level of mathematical education or expertise, in which the categories are 'Professional mathematicians etc and research students': 8 in the first group, 4 in the later one. The second category is 'Other university mathematicians': 3 in the first group, 11 in the later one. The dustbin category 'other' has 2 and 1. Recruits are also considered by age, and here the shift in the 'under 20' category, from 1 (that is Donald Michie) in the first group to 6 in the second suggests recruitment of undergraduate mathematicians or men who had taken a shortened degree course. By mid-1944 it was presumably thought that some background in mathematics would be useful and the older hands in the Newmanry felt confident that they knew what they needed to teach new recruits. This latter is confirmed by the analysis according to previous cryptographic experience. In the pre-July 1944 group, 12 have some experience, 8 have experience in Enigma work (these include Good and Wylie),[50] 3 in Fish. These numbers need some thought to interpret, since the total number of cryptographers in the group is 13, but the numbers seem to imply that only 1 member of the group had no previous experience. In the later group the figures are 3, 2 and 1 – out of a total of 16, so presumably 13 had no previous experience. (The remaining analytical criterion in the table is nationality. The first group contained 11 British and 2 Americans, the second had 13 and 3.)

The usefulness of mathematics may seem obvious, and in the context of a statistical attack it perhaps is obvious: it was Tutte, a mathematician, who set such an attack in train. However, traditionally the background for cryptanalysts had been in languages, either formally as a university degree or less formally as an aptitude for crosswords. The two chief cryptanalysts at Bletchley Park in 1939 were the head of the civilian Foreign Office group, Dillwyn Knox (1884–1943), well known as a classical scholar, and the head of the military division, John Tiltman, who had no university education but had long experience of cryptanalysis. Both were held in high regard. As it happened, Knox had been responsible for recruiting the first mathematician to join the staff at Bletchley Park: an Oxford graduate, Peter Twinn (1916–2004). Nevertheless, traditional recruitment continued. Young linguists included John Chadwick (1920–1998), who worked on Italian Enigma transmissions, having been recruited because he spoke Modern Greek fluently.[51] On the literary side, one of the Cambridge friends who wrote to Newman from Bletchley Park before Newman arrived there was the English don, and well-known literary critic, F. L. Lucas.[52]

Another traditional recruitment category was chess players, the game being seen as exemplifying logical reasoning. C. H. O'D. (Hugh) Alexander (1909–1974), of Hut 8, who had a degree in mathematics from Cambridge, was repeatedly Cambridge University chess champion and then British chess champion in 1938; Jack Good had merely been Cambridgeshire county champion. It seems likely that Alan Turing, who was not particularly good at chess, appeared suitable because he was a logician. In the event, it was obviously not lost on his colleagues that his arguments regularly took mathematical form. From about late 1941 onwards, efforts were made to recruit mathematicians for all groups.[53] Elementary mathematics had always been seen as useful – there is a practical background to those schoolroom problems about how many men it takes to dig a ditch – but now advanced, cutting-edge university mathematics was also seen as useful, if at first only in this tiny enclave of quasi-academic activity in the midst of war. The work of this group also established an association of computing with mathematics.

Consequences and Connections

To all intents and purposes, the 'universal machine' that Alan Turing described in 1936 is what we now call a computer. Depending on the historian's temperament, one can choose to see the activities of Newman's group at Bletchley Park in two opposing ways. The first is as bringing the concept closer to becoming a reality – Turing, who had a practical streak, had calculated how big a machine would need to be, and had concluded that with the means available in 1936 it would be about the size of the Albert Hall,[54] but the estimate had presumably become smaller by 1943. Alternatively, one can see the wartime activity, oriented to the narrow task of breaking Fish ciphers, as merely wasting the time of a group of exceedingly intelligent men who would have done valuable work of their own – that is assuming there had not been a war to inhibit their activities and perhaps put them in danger of their lives. These two styles of interpretation can, of course, be applied to the careers of a huge number of scientists and engineers, young and old, who found themselves involved in the war.

In the specific case of the 'universal machine' one can, however, trace some connections that seem to link the Robinson and Colossus machines with the computers built in the later 1940s in Manchester. First, there is the somewhat suggestive gathering of ex-GC&CS staff in Manchester: Newman, Good, Turing and another former member of the Newmanry with a degree in mathematics from Cambridge, David Rees (b. 1918), all worked there, in the mathematics department. And Newman had obtained a large grant from the Royal Society to make a machine capable of carrying out a variety of mathematical tasks, that is a computer. Such a machine might have seemed rather futuristic to many scientists, but the committee that approved Newman's grant included several people who had done secret work during the war and must have known about the Bletchley Park machines. To them, Newman's project no doubt seemed relatively realistic. The uses to which Newman put this grant included cooperation with the engineering department.[55] There is no doubt that the engineering skills crucial to the construction of an actual machine were contributed by Tom Kilburn, who had worked at TRE during the war, and by Professor F. C. (Freddie) Williams, but sheer contiguity makes it seem inherently implausible that the engineers never discussed their machine with the mathematicians. There was surely nothing to stop mathematicians attending an Engineering department seminar. However, none of the mathematicians was at liberty to let it be known that he already knew that a machine of this type could work, had indeed seen one working. Thanks to the secrecy surrounding the work of Bletchley Park, it was in everyone's best interests for Newman and his colleagues to play no part in the public version of the story of the origins of Baby. Now that some of the relevant wartime material has been declassified, historians are better equipped than before to arrive at a reasonable account of what happened in Manchester just after the war.

Acknowledgements

The research on which this article is based was done in connection with preparing a scholarly edition of the *General Report on Tunny with Emphasis on Statistical Methods*, working from the copy in the UK National Archives (HW 25/4, 5) and a copy held at GCHQ. We are grateful to the Royal Society for its generous support for this project. I am grateful to my co-editors, Whitfield Diffie (Sun Microsystems) and James A. Reeds (IDA, Princeton), for help understanding the technical parts of the *Report*. Furthermore, I am greatly indebted to Dr Reeds for reading an earlier version of the present article, making useful comments and patiently correcting errors.

Notes

1 See F. H. Hinsley and A. Stripp (eds), *Codebreakers: the inside story of Bletchley Park* (Oxford University Press, 1993), 20.

2 The signatories of the letter were Alan Turing, Gordon Welchman, Hugh Alexander and Stuart Milner-Barry. The letter and the reply are printed in M. Smith and R. Erskine (eds), *Action This Day* (London, New York, Toronto etc.: Bantam Press, 2001), ix–xiii. For the background, see also G. Welchman, *The Hut Six Story* (London: Allen Lane, 1982) and Hinsley and Stripp 1993.

3 Welchman 1982.

4 NARA, Record Group 457, NSA Historical Collection. Box 201, no 964. A transcription by Ralph Erskine, Philip Marks and Frode Weierud, dated February 1999, is available at http://frode.web.cern.ch/frode/crypto/Turing/index.html (last accessed 4 February 2007). On naval Enigima in the Battle of the Atlantic, see A. O. Bauer, R. Erskine and K. Herold, *Funkpeilung als alliierte Waffe gegen deutsche U-Boote 1939–1945: Wie Schwächen und Versäumnisse bei der Funkführung der U-Boote zum Ausgang der "Schlacht im Atlantik" beigetragen haben* (Rheinberg: Liebich Funk GmbH, 1997).

5 Except for the highest officials, ATS staff was generally young women, from a lower social class than the Wrens. They were always known as 'girls'.

6 The progress of the work of rebuilding a bombe is charted in a series of articles by the leader of the rebuild group, John Harper, in the magazine *Resurrection* (the newsletter of the Computer Conservation Society, which is a subgroup of the British Computer Society), between the years 1997 and 2007. See also J. V. Field, "Sigint and Automation', *IEEE Annals of the History of Computing*, 25 (No 1, January–March), 2003, 65–6 (which is an account of a lecture given by John Harper); J. Harper, 'It's Complete', *Bletchley Park Times*, Autumn 2006, 5; and B. Runciman, 'It's a bouncing baby Bombe', *IT Now* (a journal of the British Computer Society), No 32 (issue dated January 2007), 32. Further details are available at www.jharper.demon.co.uk/ bombe1.htm

7 The teleprinter had been invented in the early 1920s and had been available for one customer to dial another since the 1930s.

8 The name 'Tunny' became attached to the cipher in the summer of 1942, see *General Report on Tunny with Emphasis on Statistical Methods* [I. J. Good, D. Michie and G. Timms], 1945, 41A(a), UK National Archives (hereafter shortened to NA) HW 25/5, 297. Thrasher (the German T43) acquired its name towards the end of the war, see F. L. Bauer, 'Origins of the Fish Cypher Machines', B. J. Copeland (ed.), *Colossus: The Secrets of Bletchley Park's Codebreaking Computers* (Oxford: Oxford University Press, 2006), 411–17, esp. 417.

9 The machine, called 'Schlüsselzusatz' ('cipher attachment'), shortened to SZ – the various developed forms being given numbers of years and sometimes letters to denote specific models, to give titles such as SZ 42A – was attached to the teleprinter, so when sending or receiving messages the operators saw only plain text.
10 D. Michie, 2000, in conversation with JVF.
11 D. Michie 2000.
12 The Baudot code is given in *General Report on Tunny with Emphasis on Statistical Methods* [Good, Michie and Timms], 1945, NA HW 25/4, 3, presumably for the sake of completeness. It could have been found in standard reference works such as the *Encyclopedia Britannica*. The version in the *General Report* is reprinted in Copeland 2006, 348–9.
13 A clear account of this work is given in S. Wylie, 'Breaking Tunny and the Birth of Colossus', Smith and Erskine 2001, 317–41. Wylie, born in 1913, educated at New College, Oxford (where he took a degree in mathematics in 1934) and Princeton (PhD in mathematics 1937) worked at Bletchley Park from 1941 to 1945, at first on Enigma but from 1943 on Tunny. From 1958 to 1973 he was Chief Mathematician at GCHQ. His essay may have been informed by his having access to classified sources.
14 In the *General Report on Tunny with Emphasis on Statistical Methods* [Good, Michie and Timms], 1945, what is now called a 'bit' was called a 'character', which (confusingly) is the name that is now given to the row of five bits that make up the code for a symbol on the tape. (It seems that the use of the term 'character' in the sense we find in the *General Report* was restricted to Bletchley Park, and there was no standard term in the engineering literature.) While historians of science commonly try to adopt period vocabulary, in this case doing so seems likely to cause confusion.
15 The Tunny machine is described in detail, with some use of hindsight, that is information obtained from actually seeing machines after the end of hostilities, in *General Report on Tunny with Emphasis on Statistical Methods* [Good, Michie and Timms], 1945, sections 11A(c) and 11B(j) - 11C(c), NA HW25/4, 4–5, 10–11.
16 See the comments by Tutte, writing in 2000: William Tutte, 'My work at Bletchley Park', (Copeland 2006), Appendix 4, 352–69, esp. 367.
17 Newman's papers are now preserved in the library of St John's College, Cambridge. Correspondence relating to his recruitment to GC&CS is in box 3, folder 1. From this correspondence it appears that the person who initiated contacts between Newman and GC&CS was P. M. S. Blackett. The letter giving the date of appointment, dated 19 August 1942, is 3/1/15.
18 Newman had been the referee for Turing's paper 'On computable numbers', which describes the principles on which a computer works (A. M. Turing, 'On Computable Numbers, with an Application to the *Entscheidungsproblem', Proceedings of the London Mathematical Society*, 42 (1936), 230–65; 43 (1937), 544–6). See A. Hodges, *Alan Turing: the Enigma* (London: Vintage, 1992).
19 See *General Report on Tunny with Emphasis on Statistical Methods* [Good, Michie and Timms], 1945, 15A(a), NA HW 25/4, 33, where Tutte's method is called a '1+2-Break-in'.
20 *General Report on Tunny with Emphasis on Statistical Methods* [Good, Michie and Timms], 1945, 15B(b), NA HW 25/4, 34. This part of the Report seems to have been written by Good.
21 *General Report on Tunny with Emphasis on Statistical Methods* [Good, Michie and Timms], 1945, 74 ('Chronology''), NA HW 25/5, 458.
22 Until the true title became public, when the document was declassified in 2000, the *General Report* ... was usually known as 'The Newmanry History'. At the time of writing (March 2007) the corresponding report on the activities of the Testery, 'The Testery History', has not yet been declassified.
23 After the war Michie went up to Oxford, but instead of Classics read Medicine and then specialised in genetics. In the 1960s he became a pioneer of machine intelligence, and

was responsible for the design of prototype industrial robots. He died in a car crash on 7 July 2007.

24 Good's father was an immigrant shopkeeper in London (the family name was Gudak). Good obtained a scholarship to Jesus College, Cambridge, and after graduation went on to gain a doctorate under the supervision of the distinguished mathematician G. H. Hardy. He won a Smith's Prize (awarded for a postgraduate mathematical essay) in 1940. At the end of the war he took up a post at the University of Manchester, then from 1948 to 1959 worked in GCHQ. His continued involvement with intelligence services, in the UK and the USA, has gone together with an extremely distinguished academic career, particularly as a statistician.

25 S. L. Zabell, 'Alan Turing and the Central Limit Theorem', *American Mathematics Monthly* (1995), 483–94.

26 I. J. Good, private communication to J. A. Reeds (IDA, Princeton), 2006.

27 *General Report on Tunny with Emphasis on Statistical Methods* [Good, Michie and Timms], 1945, 21(f), last line, NA HW 25/4, 39. The theorem is named after Thomas Bayes (1702–1764).

28 Charles Eryl Wynn-Williams had worked at the Cavendish Laboratory in Cambridge in the 1930s. In 1931, he invented a device for counting alpha particles that used thyratrons, though it took the input from a conventional detector.

29 *General Report on Tunny with Emphasis on Statistical Methods* [Good, Michie and Timms], 1945, 74 ('Chronology'), NA HW 25/5, 456–63; reprinted in Copeland 2006, 338–47.

30 For the name SZ40 see note 9.

31 The text, *General Report on Tunny with Emphasis on Statistical Methods* [Good, Michie and Timms], 1945, 74, AN HW 25/5, 457, has 'March 1943' but this is presumably an error for 'March 1942'. The error is not corrected in the reprint in Copeland 2006, 339.

32 *General Report on Tunny with Emphasis on Statistical Methods* [Good, Michie and Timms], 1945, 13A(a), NA HW 25/4, 25.

33 *General Report on Tunny with Emphasis on Statistical Methods* [Good, Michie and Timms], 1945, 13B(b), NA HW 25/4, 26.

34 *General Report on Tunny with Emphasis on Statistical Methods* [Good, Michie and Timms], 1945, 52 and 54, NA HW 25/5, 328–31, 353–62.

35 *General Report on Tunny with Emphasis on Statistical Methods* [Good, Michie and Timms], 1945, 53 ('Colossus'), NA HW 25/5, 332–52.

36 *General Report on Tunny with Emphasis on Statistical Methods* [Good, Michie and Timms], 1945, 52(b)(i), NA HW 25/5, 328.

37 *General Report on Tunny with Emphasis on Statistical Methods* [Good, Michie and Timms], 1945, 51(a), NA HW 25/5, 325.

38 *General Report on Tunny with Emphasis on Statistical Methods* [Good, Michie and Timms], 1945, 52(b), NA HW 25/5, 328.

39 *General Report on Tunny with Emphasis on Statistical Methods* [Good, Michie and Timms], 1945, 52(b), NA HW 25/5, 328.

40 See B. J. Copeland, D. Bolam, H. Fensom and N. Thurlow, *Dollis Hill at War*, (Copeland 2006), 281–90.

41 Turing 1936.

42 *General Report on Tunny with Emphasis on Statistical Methods* [Good, Michie and Timms], 1945, 51(j), NA HW 25/5, 327.

43 *General Report on Tunny with Emphasis on Statistical Methods* [Good, Michie and Timms], 1945, 55(a), NA HW 25/5, 363.

44 *General Report on Tunny with Emphasis on Statistical Methods* [Good, Michie and Timms], 1945, 57(a), NA HW 25/5, 380.

45 The two types of Tunny machine are distinguished in *General Report on Tunny with Emphasis on Statistical Methods* [Good, Michie and Timms], 1945, 55J (cryptanalytic use) and 56L (decoding), NA HW 25/5, 376–9.

46 *General Report on Tunny with Emphasis on Statistical Methods* [Good, Michie and Timms], 1945, 51(k), NA HW 25/5, 327.
47 *General Report on Tunny with Emphasis on Statistical Methods* [Good, Michie and Timms], 1945, 31B, NA HW 25/5, 276.
48 *General Report on Tunny with Emphasis on Statistical Methods* [Good, Michie and Timms], 1945, 31D, NA HW 25/5, 277.
49 Timms, the elder son of a Yorkshire rope-maker, was educated at Woodhouse Grove School (Bradford). He went to Leeds University to study chemistry, but switched to mathematics after the first year and graduated with first class honours. He was then invited to study geometry at Cambridge University, under H. F. Baker at St John's College. On leaving Cambridge, in 1928, having completed his PhD work, Timms taught mathematics at St Andrews University, Scotland. After the war he joined GCHQ. See Timms' obituary *Proceedings of the Edinburgh Mathematical Society*, 26 (1983), 393–4. I am grateful to Timms' daughter, Bera MacClement, for providing information about her father.
50 On Wylie see note 13.
51 After the war Chadwick became the leading expert on Mycenaean Greek (see J. Chadwick and M. Ventris, *Documents in Mycenaean Greek* [Cambridge, 1956], and J. Chadwick, *The Decipherment of Linear B* [Cambridge, 1958]). He taught in the Classics department at the University of Cambridge. On his war work see John Chadwick, 'A Biographical Fragment: 1942–3', (Smith and Erskine, 2001), 110–26.
52 Newman papers, library of St John's College, Cambridge, 3/1/7, letter dated 27 July 1942. Lucas worked in Hut 3 (dealing with decrypts), see W. Milward, 'Life in and out of Hut 3', in: Hinsley and Stripp (1993), 17–29, esp. 24, 26.
53 See, for example, P. Hilton, 'Living with Fish: Breaking Tunny in the Newmanry and Testery', (Copeland, 2006), 189–203, esp. 189–92.
54 A substantial mid nineteenth-century building in London.
55 For Newman's work in Manchester, see D. Anderson, 'Was the Manchester Baby conceived at Bletchley Park?', University of Portsmouth Research Report Series (UoP-HC-2006-001), 2006; available at http://www.tech.port.ac.uk/staffweb/andersod/HoC/Reports/RRS_PublicViewFiles.php (last accessed 17 April 2007).

Index